MILITARY AND VETERAN ISSUES

U.S. NAVY SHIPBUILDING

PLANS, CONGRESSIONAL CONCERNS, AND QUALITY ISSUES

MILITARY AND VETERAN ISSUES

Additional books in this series can be found on Nova's website under the Series tab.

Additional E-books in this series can be found on Nova's website under the E-book tab.

MILITARY AND VETERAN ISSUES

U.S. NAVY SHIPBUILDING

PLANS, CONGRESSIONAL CONCERNS, AND QUALITY ISSUES

GLADYS CLEMENS
EDITOR

nova publishers
New York

For permission to use material from this book please contact us:
Telephone 631-231-7269; Fax 631-231-8175
Web Site: http://www.novapublishers.com

Library of Congress Cataloging-in-Publication Data

ISBN: 978-1-63117-113-0

Published by Nova Science Publishers, Inc. † New York

CONTENTS

PREFACE

This book provides background information and presents potential issues for Congress concerning the Navy's ship force-structure goals and shipbuilding plans. It also discusses, among other issues, quality problems in constructing recently delivered ships and Navy actions to improve quality and key practices employed by leading commercial ship buyers and shipbuilders to ensure quality and how these compared with Navy practices.

Chapter 1 – The Navy's proposed FY2014 budget requests funding for the procurement of 8 new battle force ships (i.e., ships that count against the Navy's goal for achieving and maintaining a fleet of 306 ships). The 8 ships include two Virginia-class attack submarines, one DDG-51 class Aegis destroyer, four Littoral Combat Ships (LCSs), and one Mobile Landing Platform/Afloat Forward Staging Base (MLP/AFSB) ship. The Navy's proposed FY2014-FY2018 five-year shipbuilding plan includes a total of 41 ships—the same number as in the Navy's FY213-FY2017 five-year shipbuilding plan, and one less than the 42 ships that the Navy planned for FY2014-FY2018 under the FY2013 budget submission.

The planned size of the Navy, the rate of Navy ship procurement, and the prospective affordability of the Navy's shipbuilding plans have been matters of concern for the congressional defense committees for the past several years. The Navy's FY2014 30-year (FY2014-FY2043) shipbuilding plan, like the Navy's previous 30-year shipbuilding plans in recent years, does not include enough ships to fully support all elements of the Navy's 306-ship goal over the long run. The Navy projects that the fleet would remain below 306 ships during most of the 30-year period, and experience shortfalls at various points in cruisers-destroyers, attack submarines, and amphibious ships.

In its October 2013 report on the cost of the FY2014 30-year shipbuilding plan, the Congressional Budget Office (CBO) estimates that the plan would cost an average of $19.3 billion per year in constant FY2013 dollars to implement, or about 15% more than the Navy estimates. CBO's estimate is about 6% higher than the Navy's estimate for the first 10 years of the plan, about 14% higher than the Navy's estimate for the second 10 years of the plan, and about 26% higher than the Navy's estimate for the final 10 years of the plan. Some of the difference between CBO's estimate and the Navy's estimate, particularly in the latter years of the plan, is due to a difference between CBO and the Navy in how to treat inflation in Navy shipbuilding.

Proposed issues for Congress in reviewing the Navy's proposed FY2014 shipbuilding budget, its proposed FY2014-FY2018 five-year shipbuilding plan, and its FY2014 30-year (FY2014- FY2043) shipbuilding plan include the following:

- the impact on Navy shipbuilding programs of the March 1, 2013, sequester on FY2013 funding and unobligated prior-year funding;
- the potential impact on Navy shipbuilding programs of a possible sequester later this year or early next year on FY2014 funding and unobligated prior-year funding;
- the potential impact on the size of the Navy of reducing DOD spending (through sequestration or regular appropriations activity) in FY2013-FY2021 to levels at or near the lower caps established in the Budget Control Act of 2011;
- the future size and structure of the Navy in light of strategic and budgetary changes;
- the sufficiency of the 30-year shipbuilding plan for achieving the Navy's goal for a 306-ship fleet; and
- the affordability of the 30-year shipbuilding plan.

Chapter 2 – The Navy expects to spend about $15 billion per year to provide its fleet with the most advanced ships to support national defense and military strategies. Problems with recently delivered ships have focused attention on quality issues.

House Report No. 112-110, accompanying the Department of Defense Appropriations Bill, 2012, mandated that GAO review the Navy's quality assurance processes for new ship construction. This report discusses, among other issues, (1) quality problems in constructing recently delivered ships and Navy actions to improve quality and (2) key practices employed by leading commercial ship buyers and shipbuilders to ensure quality and how these compared with Navy practices.

GAO analyzed Navy data on ship quality from 2006 to May 2013 and spoke with Navy officials and shipbuilders. GAO also reviewed deficiency data for commercial ships and spoke with buyers and builders.

In: U.S. Navy Shipbuilding
Editor: Gladys Clemens
ISBN: 978-1-63117-113-0

Chapter 1

NAVY FORCE STRUCTURE AND SHIPBUILDING PLANS: BACKGROUND AND ISSUES FOR CONGRESS*

Ronald O'Rourke

SUMMARY

The Navy's proposed FY2014 budget requests funding for the procurement of 8 new battle force ships (i.e., ships that count against the Navy's goal for achieving and maintaining a fleet of 306 ships). The 8 ships include two Virginia-class attack submarines, one DDG-51 class Aegis destroyer, four Littoral Combat Ships (LCSs), and one Mobile Landing Platform/Afloat Forward Staging Base (MLP/AFSB) ship. The Navy's proposed FY2014-FY2018 five-year shipbuilding plan includes a total of 41 ships—the same number as in the Navy's FY213-FY2017 five-year shipbuilding plan, and one less than the 42 ships that the Navy planned for FY2014-FY2018 under the FY2013 budget submission.

The planned size of the Navy, the rate of Navy ship procurement, and the prospective affordability of the Navy's shipbuilding plans have been matters of concern for the congressional defense committees for the past several years. The Navy's FY2014 30-year (FY2014-FY2043) shipbuilding plan, like the Navy's previous 30-year shipbuilding plans in recent years, does not include enough ships to fully support all elements of the Navy's 306-ship goal over the long run. The Navy projects that the fleet would remain below 306 ships during most of the 30-year period, and experience shortfalls at various points in cruisers-destroyers, attack submarines, and amphibious ships.

In its October 2013 report on the cost of the FY2014 30-year shipbuilding plan, the Congressional Budget Office (CBO) estimates that the plan would cost an average of $19.3 billion per year in constant FY2013 dollars to implement, or about 15% more than the Navy estimates. CBO's estimate is about 6% higher than the Navy's estimate for the first 10 years of the plan, about 14% higher than the Navy's estimate for the second 10 years of the plan, and about 26% higher than the Navy's estimate for the final 10 years of the plan. Some of the

* This is an edited, reformatted and augmented version of the Congressional Research Service Publication, CRS Report for Congress RL32665, dated November 8, 2013.

difference between CBO's estimate and the Navy's estimate, particularly in the latter years of the plan, is due to a difference between CBO and the Navy in how to treat inflation in Navy shipbuilding.

Proposed issues for Congress in reviewing the Navy's proposed FY2014 shipbuilding budget, its proposed FY2014-FY2018 five-year shipbuilding plan, and its FY2014 30-year (FY2014- FY2043) shipbuilding plan include the following:

- the impact on Navy shipbuilding programs of the March 1, 2013, sequester on FY2013 funding and unobligated prior-year funding;
- the potential impact on Navy shipbuilding programs of a possible sequester later this year or early next year on FY2014 funding and unobligated prior-year funding;
- the potential impact on the size of the Navy of reducing DOD spending (through sequestration or regular appropriations activity) in FY2013-FY2021 to levels at or near the lower caps established in the Budget Control Act of 2011;
- the future size and structure of the Navy in light of strategic and budgetary changes;
- the sufficiency of the 30-year shipbuilding plan for achieving the Navy's goal for a 306-ship fleet; and
- the affordability of the 30-year shipbuilding plan.

INTRODUCTION

This report provides background information and presents potential issues for Congress concerning the Navy's ship force-structure goals and shipbuilding plans. The planned size of the Navy, the rate of Navy ship procurement, and the prospective affordability of the Navy's shipbuilding plans have been matters of concern for the congressional defense committees for the past several years. Decisions that Congress makes on Navy shipbuilding programs can substantially affect Navy capabilities and funding requirements, and the U.S. shipbuilding industrial base.

BACKGROUND

Navy's Ship Force Structure Goal

January 2013 Goal for Fleet of 306 Ships

On January 31, 2013, in response to Section 1015 of the FY2013 National Defense Authorization Act (H.R. 4310/P.L. 112-239 of January 2, 2013), the Navy submitted to Congress a report presenting a goal for achieving and maintaining a fleet of 306 ships, consisting of certain types and quantities of ships.[1] The goal for a 306-ship fleet is the result of a force structure assessment (FSA) that the Navy completed in 2012.

306-Ship Goal Reflects 2012 Strategic Guidance and Projected DOD Spending Shown in FY2013 and FY2014 Budget Submissions

The 2012 FSA and the resulting 306-ship plan reflect the defense strategic guidance document that the Administration presented in January 2012[2] and the associated projected levels of Department of Defense (DOD) spending shown in the FY2013 and FY2014 budget submissions. DOD officials have stated that if planned levels of DOD spending are reduced below what is shown in these budget submissions, the defense strategy set forth in the January 2012 strategic guidance document might need to be changed. Such a change, Navy officials have indicated, could lead to the replacement of the 306-ship plan of January 2013 with a new plan.

Goal for Fleet of 306 Ships Compared to Earlier Goals

Table 1 compares the 306-ship goal to earlier Navy ship force structure plans.

Navy's Five-Year and 30-Year Shipbuilding Plans

Five-Year (FY2014-FY2018) Shipbuilding Plan

Table 2 shows the Navy's FY2014 five-year (FY2013-FY2018) shipbuilding plan.

Observations that can be made about the Navy's proposed five-year (FY2014-FY2018) shipbuilding plan include the following:

- **Total of 41 ships—about the same as last year.** The Navy's proposed FY2014-FY2018 five-year shipbuilding plan includes a total of 41 ships—the same number as in the Navy's FY213-FY2017 five-year shipbuilding plan, and one less than the 42 ships that the Navy planned for FY2014-FY2018 under the FY2013 budget submission.
- **Average of 8.2 ships per year.** The FY2013-FY2017 plan includes an average of 8.2 battle force ships per year. The steady-state replacement rate for a fleet of 306 ships with an average service life of 35 years is about 8.7 ships per year. In light of how the average shipbuilding rate since FY1993 has been substantially below 8.7 ships per year (see **Appendix D**), shipbuilding supporters for some time have wanted to increase the shipbuilding rate to a steady rate of 10 or more battle force ships per year.
- **Second Virginia-class submarine added in FY2014.** Compared to the FY2013-FY2017 five-year shipbuilding plan, the FY2014-FY2018 five-year shipbuilding plan adds a second Virginia-class attack submarine in FY2014. This follows through on Congress's action, in marking up the FY2013 budget, to provide adv ance procurement funding in FY2013 for a second Virginia-class boat in FY2014.

Table 1. Current 306 Ship Force Structure Goal Compared to Earlier Goals

Ship type	306- ship plan of January 2013	~310- 316 ship plan of March 2012	Revised 313-ship plan of September 2011	Changes to February 2006 313-ship plan announced through mid-2011	February 2006 Navy plan for 313-ship fleet	Early-2005 Navy plan for fleet of 260-325 ships		2002- 2004 Navy plan for 375-ship Navy[a]	2001 QDR plan for 310- ship Navy
						260-ships	325-ships		
Ballistic missile submarines (SSBNs)	12[b]	12-14[b]	12[b]	12[b]	14	14	14	14	14
Cruise missile submarines (SSGNs)	0[c]	0-4[c]	4[c]	0[c]	4	4	4	4	2 or 4[d]
Attack submarines (SSNs)	48	~48	48	48	48	37	41	55	55
Aircraft carriers	11[e]	11[e]	11[e]	11[e]	11f	10	11	12	12
Cruisers and destroyers	88	~90	94	94[g]	88	67	92	104	116
Frigates	0	0	0	0	0	0	0	0	
Littoral Combat Ships (LCSs)	52	~55	55	55	55	63	82	56	0
Amphibious ships	33	~32	33	33[h]	31	17	24	37	36
MPF(F) shipsi	0[j]	0[j]	0[j]	0[j]	12[i]	14[i]	20[i]	0[i]	0[i]
Combat logistics (resupply) ships	29	~29	30	30	30	24	26	42	34
Dedicated mine warfare ships	0	0	0	0	0	0	0	26[k]	16
Joint High Speed Vessels (JHSVs)	10[l]	10[l]	10[l]	21[l]	3	0	0	0	0
Otherm	23	~23	16	24n	17	10	11	25	25
Total battle force ships	**306**	**~310-316**	**313**	**328**	**313**	**260**	**325**	**375**	**310 or 312**

Sources: Table prepared by CRS based on U.S. Navy data.

Note: QDR is Quadrennial Defense Review. The "~" symbol means approximately and signals that the number in question may be refined as a result of the Naval Force Structure Assessment currently in progress.

a. Initial composition. Composition was subsequently modified.

b. The Navy plans to replace the 14 current Ohio-class SSBNs with a new class of 12 next-generation SSBNs. For further discussion, see CRS Report R41129, *Navy Ohio Replacement (SSBN[X]) Ballistic Missile Submarine Program: Background and Issues for Congress*, by Ronald O'Rourke.

c. Although the Navy plans to continue operating its four SSGNs until they reach retirement age in the late 2020s, the Navy does not plan to replace these ships when they retire. This situation can be expressed in a table like this one with either a 4 or a zero.

d. The report on the 2001 QDR did not mention a specific figure for SSGNs. The Administration's proposed FY2001 DOD budget requested funding to support the conversion of two available Trident SSBNs into SSGNs, and the retirement of two other Trident SSBNs. Congress, in marking up this request, supported a plan to convert all four available SSBNs into SSGNs.

e. With congressional approval, the goal has been temporarily be reduced to 10 carriers for the period between the retirement of the carrier *Enterprise* (CVN-65) in December 2012 and entry into service of the carrier *Gerald R. Ford* (CVN-78), currently scheduled for September 2015.

f. For a time, the Navy characterized the goal as 11 carriers in the nearer term, and eventually 12 carriers.

g. The 94-ship goal was announced by the Navy in an April 2011 report to Congress on naval force structure and missile defense.

h. The Navy acknowledged that meeting a requirement for being able to lift the assault echelons of 2.0 Marine Expeditionary Brigades (MEBs) would require a minimum of 33 amphibious ships rather than the 31 ships shown in the February 2006 plan. For further discussion, see CRS Report RL34476, *Navy LPD-17 Amphibious Ship Procurement: Background, Issues, and Options for Congress*, by Ronald O'Rourke.

i. Today's Maritime Prepositioning Force (MPF) ships are intended primarily to support Marine Corps operations ashore, rather than Navy combat operations, and thus are not counted as Navy battle force ships. The planned MPF (Future) ships, however, would have contributed to Navy combat capabilities (for example, by supporting Navy aircraft operations). For this reason, the ships in the planned MPF(F) squadron were counted by the Navy as battle force ships. The planned MPF(F) squadron was subsequently restructured into a different set of initiatives for enhancing the existing MPF squadrons; the Navy no longer plans to acquire an MPF(F) squadron.

j. The Navy no longer plans to acquire an MPF(F) squadron. The Navy, however, has procured or plans to procure some of the ships that were previously planned for the squadron—specifically, TAKE-1 class cargo ships, and Mobile Landing Platform (MLP)/Afloat Forward Staging Base (AFSB) ships. These ships are included in the total shown for "Other" ships.

k. The figure of 26 dedicated mine warfare ships included 10 ships maintained in a reduced mobilization status called Mobilization Category B. Ships in this status are not readily deployable and thus do not count as battle force ships. The 375-ship proposal thus implied transferring these 10 ships to a higher readiness status.

l. Totals shown include 5 ships transferred from the Army to the Navy and operated by the Navy primarily for the performance of Army missions.

m. This category includes, among other things, command ships and support ships.

n. The increase in this category from 17 ships under the February 2006 313-ship plan to 24 ships under the apparent 328-ship goal included the addition of one TAGOS ocean surveillance ship and the transfer into this category of six ships—three modified TAKE-1 class cargo ships, and three Mobile Landing Platform (MLP) ships—that were previously intended for the planned (but now canceled) MPF(F) squadron.

Table 2. Navy FY2014 Five-Year (FY2014-FY2018) Shipbuilding Plan (Battle force ships—i.e., ships that count against 306-ship goal)

Ship type	FY14	FY15	FY16	FY17	FY18	Total
Ford (CVN-78) class aircraft carrier	0	0	0	0	1	1
Virginia (SSN-774) class attack submarine	2	2	2	2	2	10
Arleigh Burke (DDG-51) class destroyer	1	2	2	2	2	9
Littoral Combat Ship (LCS)	4	4	2	2	2	14
LHA(R) amphibious assault ship	0	0	0	1	0	1
Fleet tug (TATF)	0	0	0	2	1	3
Mobile Landing Platform (MLP)/Afloat Forward Staging Base (AFSB)	1	0	0	0	0	1
TAO(X) oiler	0	0	1	0	1	2
TOTAL	**8**	**8**	**7**	**9**	**9**	**41**

Source: FY2014 Navy budget submission.

Notes: The MLP/AFSB is a variant of the MLP with additional features permitting it to serve in the role of an AFSB. The Navy proposes to fund the TATFs and TAO(X)s through the National Defense Sealift Fund (NDSF) and the other ships through the Navy's shipbuilding account, known formally as the Shipbuilding and Conversion, Navy (SCN) appropriation account.

- **One DDG-51 destroyer in FY2014.** The FY2014-FY2018 five-year shipbuilding plan, like the FY2013-FY2017 five-year shipbuilding plan, includes one DDG-51 destroyer in FY2014. During Congress's review of the Navy's FY2013 budget submission, there was interest in Congress in adding a second DDG-51 destroyer to FY2014. In final action, Congress instead procured an additional DDG-51 destroyer in FY2013 (increasing the number of DDG-51s procured in FY2013 to three, compared to the two that were requested for FY2013). The third DDG-51 procured in FY2013 might be viewed as the equivalent of the second DDG-51 that supporters wanted to add to FY2014.
- **Start of LX(R) amphibious ship procurement deferred beyond FY2018.** The proposed FY2014-FY2018 five-year shipbuilding plan defers the procurement of the first LX(R) amphibious ship from FY2018 to a later fiscal year. The FY2013-FY2017 five-year shipbuilding plan had deferred the ship from FY2017 to FY2018. LX(R)s (previously designated LSD[X]s) are to replace aging LSD- 41/49 class amphibious ships.

30-Year (FY2014-FY2043) Shipbuilding Plan

Table 3 shows the Navy's FY2014 30-year (FY2014-FY2043) shipbuilding plan, the tables for which were submitted to Congress on April 22, 2012.

In devising a 30-year shipbuilding plan to move the Navy toward its ship force-structure goal, key assumptions and planning factors include but are not limited to the following:

- ship service lives;
- estimated ship procurement costs;
- projected shipbuilding funding levels; and
- industrial-base considerations.

The Navy states that

> This [FY2014] 30-year shipbuilding plan is based on these key assumptions:
>
> — [A] Battle Force inventory of the "2012 Navy FSA [Force Structure Assessment]" [i.e., of 306 ships] will remain the objective of this plan.
>
> — In the near term [i.e., FY2014-FY2023], the Annual budget for Navy shipbuilding will be sustained at the levels of the FY14 President's Budget (PB14) [i.e., the Navy's proposed FY2014 budget] through the Future Year Defense Plan (FYDP) [i.e., during FY2014- Fy2018]. In the med-term [i.e., FY2024-FY2033], [the] annual budget will remain at appropriate (higher) levels; and in the far term [i.e., FY2034-FY2043], [it will] be sustained at appropriate levels (slightly higher than [the] current historical average).
>
> — All battle force ships serve to the end of the planned or extended lives.
>
> — The DoN [Department of the Navy] will continue to acquire and build ships in the most affordable manner.[3]

Table 3. Navy FY2014 30-Year (FY2014-FY2043) Shipbuilding Plan

FY	CVN	LSC	SSC	SSN	SSBN	AWS	CLF	Supt	Total
14		1	4	2				1	8
15		2	4	2					8
16		2	2	2			1		7
17		2	2	2		1		2	9
18	1	2	2	2			1	1	9
19		2	3	2		1	1	1	10
20		2	3	2			1	2	10
21		2	3	2	1	1	1		10
22		3	3	2			1	2	11
23	1	3	3	2		1	1	3	14
24		2	3	1	1	1	1	2	11
25		3	3	2		1	1	1	11
26		2	1	1	1		1		6
27		3		2	1	1	1		8
28	1	3		1	1	2	1	1	10
29		3		1	1	1	1	1	8
30		2	1	1	1	1	1	2	9
31		2		2	1	1	1	2	9
32		2	1	1	1	2	1	3	11
33	1	2		1	1	1	1	2	9
34		2	1	1	1			2	7
35		2	1	1	1				5
36		2		1		1			4
37		2	4	2					8
38	1	3	4	2					10
39		3	4	1					8
40		3	4	2		2			11
41		3	4	1					8
42		3	3	2		1			9
43	1	2	3	1			1		8

Source: FY2014 30-year (FY2014-FY2043) shipbuilding plan.

Key: **FY** = Fiscal Year; **CVN** = aircraft carriers; **LSC** = surface combatants (i.e., cruisers and destroyers); **SSC** = small surface combatants (i.e., Littoral Combat Ships [LCSs]); **SSN** = attack submarines; **SSGN** = cruise missile submarines; **SSBN** = ballistic missile submarines; **AWS** = amphibious warfare ships; **CLF** = combat logistics force (i.e., resupply) ships; **Supt** = support ships.

Navy's Projected Force Levels Under 30-Year Shipbuilding Plan

Table 4 shows the Navy's projection of force levels for FY2014-FY2043 that would result from implementing the FY2014 30-year (FY2014-FY2043) shipbuilding plan shown in **Table 3**.

Table 4. Projected Force Levels Resulting from FY2014 30-Year (FY2014-FY2043) Shipbuilding Plan

	CVN	LSC	SSC	SSN	SSGN	SSBN	AWS	CLF	Supt	Total
306 ship plan	**11**	**88**	**52**	**48**	**0**	**12**	**33**	**29**	**33**	**306**
FY14	10	85	26	55	4	14	31	31	26	**282**
FY15	10	78	23	55	4	14	28	29	29	**270**
FY16	11	82	27	53	4	14	29	29	31	**280**
FY17	11	83	29	50	4	14	30	29	33	**283**
FY18	11	84	33	52	4	14	31	29	33	**291**
FY19	11	86	38	52	4	14	31	29	35	**300**
FY20	11	87	37	49	4	14	31	29	33	**295**
FY21	11	88	37	49	4	14	31	29	33	**296**
FY22	12	87	39	48	4	14	31	29	33	**297**
FY23	12	87	38	48	4	14	31	29	34	**297**
FY24	12	89	40	48	4	14	32	29	34	**302**
FY25	11	88	42	47	4	14	34	29	34	**303**
FY26	11	89	45	46	2	14	33	29	33	**302**
FY27	11	91	48	45	1	13	33	29	33	**304**
FY28	11	90	51	43	0	12	33	29	33	**302**
FY29	11	88	52	42	0	11	33	29	33	**299**
FY30	11	86	52	43	0	11	32	29	33	**297**
FY31	11	82	52	44	0	11	32	29	33	**294**
FY32	11	81	52	45	0	10	32	29	34	**294**
FY33	11	81	52	46	0	10	33	29	34	**296**
FY34	11	80	52	47	0	10	34	29	34	**297**
FY35	11	82	52	48	0	10	33	29	34	**299**
FY36	11	84	52	50	0	10	33	29	34	**303**
FY37	11	86	52	51	0	10	34	29	33	**306**
FY38	11	88	52	50	0	10	33	29	34	**307**
FY39	11	90	52	50	0	10	33	29	33	**308**
FY40	10	90	52	50	0	10	32	29	33	**308**
FY41	10	90	52	49	0	11	33	29	33	**307**
FY42	10	88	52	51	0	12	32	29	33	**307**
FY43	10	88	52	51	0	12	31	29	33	**306**

Source: FY2014 30-year (FY2014-FY2043) shipbuilding plan.

Note: Figures for support ships include five JHSVs transferred from the Army to the Navy and operated by the Navy primarily for the performance of Army missions.

Key: **FY** = Fiscal Year; **CVN** = aircraft carriers; **LSC** = surface combatants (i.e., cruisers and destroyers); **SSC** = small surface combatants (i.e., frigates, Littoral Combat Ships [LCSs], and mine warfare ships); **SSN** = attack submarines; **SSGN** = cruise missile submarines; **SSBN** = ballistic missile submarines; **AWS** = amphibious warfare ships; **CLF** = combat logistics force (i.e., resupply) ships; **Supt** = support ships.

Observations that can be made about the Navy's FY2014 30-year (FY2014-FY2043) shipbuilding plan and resulting projected force levels included the following:

- **Total of 266 ships; average of about 8.9 per year.** The plan includes a total of 266 ships to be procured, two less than the number in the FY2013 30-year (FY2013-FY2042) shipbuilding plan. The total of 266 ships equates to an average of about 8.9 ships per year, which is slightly higher than the approximate average procurement rate (sometimes called the steady-state replacement rate) of about 8.7 ships per year that would be needed over the long run to achieve and maintain a fleet of 306 ships, assuming an average life of 35 years for Navy ships.
- **Projected fleet remains below 306 ships.** Although the FY2014 30-year plan includes an average of about 8.9 ships per year, the FY2014 30-year plan, like previous 30-year plans, results in a fleet that does not fully support all elements of the Navy's ship force structure goal over the 30-year period. The distribution of the 266 ships over the 30-year period, combined with the ages of the Navy's existing ships, results in a projected fleet that would remain below 306 ships during most of the 30-year period and experience shortfalls at various points in cruisers-destroyers, attack submarines, and amphibious ships.
- **Ballistic missile submarine force to be reduced temporarily to 10 boats.** As a result of a decision in the FY2013 budget to defer the scheduled procurement of the first Ohio replacement (SSBN[X]) ballistic missile submarine by two years, from FY2019 to FY2021, the ballistic missile submarine force is projected to drop to a total of 10 or 11 boats—one or two boats below the 12-boat SSBN force-level goal—during the period FY2029-FY2041. The Navy says this reduction is acceptable for meeting current strategic nuclear deterrence mission requirements, because none of the 10 or 11 boats during these years will be encumbered by long-term maintenance.[4]
- **Seven CG-47 class cruisers and two LSD-41 class ships again proposed for early retirement.** The Navy's FY2013 budget submission proposed the early retirements in FY2013 and FY2014 of seven CG-47 class Aegis cruisers and two LSD-41 class amphibious ships. Congress, in acting on the Navy's proposed FY2013 budget, did not accept this proposal, and instead instructed the Navy to keep these nine ships in service. Section 8103 of the FY2013 DOD appropriations act (Division C of H.R. 933/P.L. 113-6 of March 26, 2013, the Consolidated and Further Continuing Appropriations Act, 2013) established a Ship Modernization, Operations and Sustainment Fund to fund the continued operation and support of these nine ships in FY2013 and FY2014. The Navy's FY2014 30-year shipbuilding plan again proposes the early retirements of seven CG-47 class Aegis cruisers and two LSD-41 class amphibious ships, with the retirement to occur in FY2015. The exact cruisers and amphibious ships to be retired under the Navy's FY2014 submission might differ from the exact cruisers and amphibious ships to be retired under the Navy's FY2013 budget submission, but the general idea of having early retirements for seven CG-47 class cruisers and two LSD-41 class amphibious appears to be the same, except that the early retirements would now occur in FY2015 (i.e., one year beyond the two-year window funded by Section 8103).

Comparison of First 10 Years of 30-Year Plans

Table 5 and **Table 6** below show the first 10 years of planned annual ship procurement quantities and projected Navy force sizes in 30-year shipbuilding plans dating back to the first such plan, which was submitted in 2000 in conjunction with the FY2001 budget. By reading vertically down each column, one can see how the ship procurement quantity or Navy force size projected for a given fiscal year changed as that year drew closer to becoming the current budget year.

OVERSIGHT ISSUES FOR CONGRESS FOR FY2014

Impact of March 1, 2013, Sequester on FY2013 Funding

June 2013 DOD Report on March 1, 2013, Sequester

One issue for Congress concerns the impact on Navy shipbuilding programs of the March 1, 2013, sequester on FY2013 funding and unobligated prior-year funding. DOD's June 2013 report to Congress on the March 1, 2013, sequester states that the sequester reduced FY2013 and unobligated prior-year funding in the Navy's shipbuilding account (known formally as the Shipbuilding and Conversion, Navy, or SCN, appropriation account) by $1,752.7 million, or about 7.6%, including

- $541.2 million in the DDG-51 destroyer program;
- $492.3 million in the Virginia class attack submarine program;
- $184.2 million in the LCS program;
- $176.3 in the program for performing mid-life nuclear refueling overhauls (called refueling complex overhauls, or RCOHs) on existing Nimitz (CVN-68) class aircraft carriers;
- $74.5 million in the program for building new CVN-78 class aircraft carriers;
- $71.1 million for the LHA Replacement amphibious assault ship program;
- $70.3 million for the DDG-1000 destroyer program;
- $58.8 million in the LPD-17 amphibious ship program;
- $25.7 million for the Moored Training Ship (MTS) program to convert an older attack submarine into a moored training platform;
- $24.7 million for ship outfitting costs;
- $21.4 million for the JHSV program;
- $10.0 million for the LCAC SLEP (air cushioned landing craft service life extension program); and
- amounts of less than $1 million each from a few other SCN-funded programs.[5]

The first two items listed above account for $1,033.5 million, or about 59.0%, of the total amount sequestered from the SCN account. The first four items listed above account for $1,394.0 million, or about 79.5%, of the total amount sequestered from the SCN account.

Table 5. Ship Procurement Quantities in First 10 Years of 30-Year Shipbuilding Plans
Years shown are fiscal years

FY of 30 - year plan (year submitted)	01	02	03	04	05	06	07	08	09	10	11	12	13	14	15	16	17	18	19	20	21	22	23
FY01 plan (2000)	8	8	8	8	7	5	6	6	6	7													
FY02 plan (2001)		6	n/a	n/a	n/a	n/a	n/a	n/a	n/a	n/a	n/a												
FY03 plan (2002)			5	5	7	7	11	n/a	n/a	n/a	n/a	n/a											
FY04 plan (2003)				7	8	7	7	9	14	15	13	14	15										
FY05 plan (2004)					9	6	8	9	17	14	15	14	16	15									
FY06 plan (2005)						4	7	7	9	10	12	n/a	n/a	n/a	n/a								
FY07 plan (2006)							7	7	11	12	14	13	12	11	11	10							
FY08 plan (2007)								7	11	12	13	12	12	10	12	11	6						
FY09 plan (2008)									7	8	8	12	12	13	13	12	12	13					
FY10 plan (2009)										8	n/a	n/a	n/a	n/a	n/a	n/a	n/a	n/a	n/a				
FY11 plan (2010)											9	8	12	9	12	9	12	9	13	9			
FY12 plan (2011)												10	13	11	12	9	12	10	12	8	9		
FY13 plan (2012)													10	7	8	9	7	11	8	12	9	12	
FY14 plan (2013)														8	8	7	9	9	10	10	10	11	14

Source: Navy 30-year shipbuilding plans supplemented by annual Navy budget submissions (including 5-year shipbuilding plans) for fiscal years shown. **n/a** means not available—see notes below.

Notes: The FY2001 30-year plan submitted in 2000 was submitted under a one-time-only legislative provision, Section 1013 of the FY2000 National Defense Authorization Act (S. 1059/P.L. 106-65 of October 5, 1999). No provision required DOD to submit a 30-year shipbuilding plan in 2001 or 2002, when Congress considered DOD's proposed FY2002 and FY2003 DOD budgets. (In addition, no FYDP was submitted in 2001, the first year of the George W. Bush Administration.) Section 1022 of the FY2003 Bob Stump National Defense Authorization Act (H.R. 4546/P.L. 107-314 of December 2, 2002) created a requirement to submit a 30-year shipbuilding plan each year, in conjunction with each year's defense budget. This provision was codified at 10 U.S.C. 231. The first 30-year plan submitted under this provision was the one submitted in 2003, in conjunction with the proposed FY2004 DOD budget. For the next several years, 30-year shipbuilding plans were submitted each year, in conjunction with each year's proposed DOD budget. An exception occurred in 2009, the first year of the Obama Administration, when DOD submitted a proposed budget for FY2010 with no accompanying FYDP or 30-year Navy shipbuilding plan. Section 1023 of the FY2011 Ike Skelton National Defense Authorization Act (H.R. 6523/P.L. 111-383 of January 7, 2011) amended 10 U.S.C. 231 to require DOD to submit a 30-year shipbuilding plan once every four years, in the same year that DOD submits a Quadrennial Defense Review

(QDR). Consistent with Section 1023, DOD did not submit a new 30-year shipbuilding plan at the time that it submitted the proposed FY2012 DOD budget. At the request of the House Armed Services Committee, the Navy submitted the FY2012 30-year (FY2012-FY2041) shipbuilding plan in late-May 2011. Section 1011 of the FY2012 National Defense Authorization Act (H.R. 1540/P.L. 112-81 of December 31, 2011) amended 10 U.S.C. 231 to reinstate the requirement to submit a 30-year shipbuilding plan each year, in conjunction with each year's defense budget.

Table 6. Projected Navy Force Sizes in First 10 years of 30-Year Shipbuilding Plans
Years shown are fiscal years

FY of 30-year plan (year submitted)	01	02	03	04	05	06	07	08	09	10	11	12	13	14	15	16	17	18	19	20	21	22	23
FY01 plan(2000)	316	315	313	313	313	311	311	304	305	305													
FY02 plan (2001)		316	n/a	n/a	n/a	n/a	n/a	n/a	n/a	n/a	n/a												
FY03 plan (2002)			314	n/a	n/a	n/a	n/a	n/a	n/a	n/a	n/a	n/a											
FY04 plan (2003)				292	292	291	296	301	305	308	313	317	321										
FY05 plan (2004)					290	290	298	303	308	307	314	320	328	326									
FY06 plan (2005)						289	293	297	301	301	306	n/a	n/a	305	n/a								
FY07 plan (2006)							285	294	299	301	306	315	317	315	314	317							
FY08 plan (2007)								286	289	293	302	310	311	307	311	314	322						
FY09 plan (2008)									286	287	289	290	293	287	288	291	301	309					
FY10 plan (2009)										287	n/a	n/a	n/a	n/a	n/a	n/a	n/a	n/a	n/a				
FY11 plan (2010)											284	287	287	285	285	292	298	305	311	315			
FY12 plan (2011)												290	287	286	286	297	301	311	316	322	324		
FY13 plan (2012)													285	279	276	284	285	292	300	295	296	298	
FY14 plan (2013)														282	270	280	283	291	300	295	296	297	297

Source: Navy 30-year shipbuilding plans supplemented by annual Navy budget submissions (including 5-year shipbuilding plans) for fiscal years shown. **n/a** means not available—see notes below.

Notes: The FY2001 30-year plan submitted in 2000 was submitted under a one-time-only legislative provision, Section 1013 of the FY2000 National Defense Authorization Act (S. 1059/P.L. 106-65 of October 5, 1999). No provision required DOD to submit a 30-year shipbuilding plan in 2001 or 2002, when Congress considered DOD's proposed FY2002 and FY2003 DOD budgets. Section 1022 of the FY2003 Bob Stump National Defense Authorization Act (H.R. 4546/P.L. 107-314 of December 2, 2002) created a requirement to submit a 30-year shipbuilding plan each year, in conjunction with each year's defense budget. This provision was codified at 10 U.S.C. 231. The first 30-year plan submitted under this provision was the one submitted in 2003, in conjunction with the proposed FY2004 DOD budget. For the next several years, 30-year shipbuilding plans were submitted each year, in conjunction with each year's proposed DOD budget. An exception occurred in 2009, the first year of the Obama Administration, when DOD submitted a proposed budget for FY2010 with no accompanying FYDP or 30-year Navy shipbuilding plan. The FY2006 plan included data for only selected years beyond FY2011.

Section 1023 of the FY2011 Ike Skelton National Defense Authorization Act (H.R. 6523/P.L. 111-383 of January 7, 2011) amended 10 U.S.C. 231 to require DOD to submit a 30-year shipbuilding plan once every four years, in the same year that DOD submits a Quadrennial Defense Review (QDR). Consistent with Section 1023, DOD did not submit a new 30-year shipbuilding plan at the time that it submitted the proposed FY2012 DOD budget. At the request of the House Armed Services Committee, the Navy submitted the FY2012 30-year (FY2012-FY2041) shipbuilding plan in late-May 2011. Section 1011 of the FY2012 National Defense Authorization Act (H.R. 1540/P.L. 112-81 of December 31, 2011) amended 10 U.S.C. 231 to reinstate the requirement to submit a 30-year shipbuilding plan each year, in conjunction with each year's defense budget.

The impact of the March 1, 2013, sequester on an individual Navy shipbuilding program will depend on the particular circumstances of that program, including, among other things, the impact of the March 1, 2013, sequester on funding for research and development efforts supporting that program.

Executability of Third DDG-51 Funded in FY2013

A key question regarding the impact of the sequester on FY2013 funding concerns the executability (i.e., on the Navy's ability to go ahead with the construction) of the third DDG-51 destroyer that was funded in FY2013. This ship is to be the 10th ship in a DDG-51 multiyear procurement (MYP) contract for FY2013-FY2017 that was awarded in early June 2013.

At a May 8, 2013, hearing on Navy shipbuilding programs before the Seapower subcommittee of the Senate Armed Services Committee, Sean Stackley, the Assistant Secretary of the Navy for Research, Development and Acquisition (i.e., the Navy's acquisition executive), testified that as a result of the March 1, 2013, sequester, there is an approximately $300 million funding shortfall in the DDG-51 program that creates a problem for executing the third DDG-51 that was funded in FY2013.[6] Stackley stated that the Navy is ready to work with Congress to address this issue, and that the ship is treated as an option in the DDG-51 MYP contract.[7]

The suggestion from the Navy's testimony is that if Congress were to address the funding shortfall for the DDG-51 program this year, either through its action on the Navy's proposed FY2014 budget or in some other way, the ship could be added to the MYP contract following the award of the contract by exercising the contract's option for the ship. If that were to happen this year, it would not substantially affect the schedule for building this ship, because the Navy from the start has anticipated building the ship on a schedule consistent with what would be expected for a ship funded in FY2014. In other words, for construction scheduling purposes, the Navy from the start has anticipated treating the ship like a "second FY2014 DDG-51" rather than a third FY2013 DDG-51.

November 7, 2013, Navy Testimony

At a November, 7, 2013, hearing before the Senate Armed Services Committee on the impact of sequestration on the national defense, Admiral Jonathan Greenert testified, in a statement similar to the one he used at the September 18, 2013, hearing discussed below, that

> Looking at the nearer term, the FY 2013 sequestration reductions compelled us to reduce our afloat and ashore operations and created a significant shore maintenance backlog. However, the effects were barely manageable because we received authorization to reprogram funds into appropriate maintenance accounts, and we were able to use prior-year investment balances to mitigate reductions to investment programs. Impact to Navy programs, caused by the combination of a continuing resolution and sequestration, included:
>
> - Cancelled five ship deployments.
> - Delayed deployment of USS HARRY S TRUMAN strike group by six months.
> - Planned inactivation, instead of repairing, USS MIAMI due to rising cost and inadequate maintenance funds.

- Reduced facilities restoration and modernization by about 30%.
- Furloughed DON civilian employees for 6 days, which, combined with a hiring freeze, reduced our maintenance and sustainment capacity by taking away logisticians, comptrollers, engineers, contracting officers, and planners.
- Reduced base operations, including port and airfield operations, by about 20%.
- Cancelled the Blue Angels' season and most non-essential port visits for Fleet Weeks.[8]

October 23, 2013, Navy Testimony

At an October 23, 2013, hearing before the Tactical Air and Land Forces subcommittee of the House Armed Services Committee on the impacts of a continuing resolution (CR) and sequestration on acquisition and modernization, Department of the Navy officials testified that

> ... FY 2013 sequestration reduced the DON top-line by approximately $11 billion and impacted our readiness, operations and procurement. The effects were addressed by curtailing operations, deferring maintenance, depleting unobligated prior-year balances in our investment accounts, deferring costs to future year budgets, and employing the limited transfer authority provided to the DON. These measures which mitigated the immediate impacts were insufficient to bear the full weight of sequestration, resulting in delays to development schedules and reductions to procurement quantities. The net effects of these deferrals, delays, and program cuts are added bills and increased costs in FY 2014 and beyond.
>
> In support of our warfighters, the DON prioritized readiness for deployed and next to deploy forces. By this action, the Navy and Marine Corps were able to preserve the capability and capacity to meet the highest priority Combatant Commander demands, including crisis response and theater security cooperation, although participation levels were reduced in some cases. However, coupled with the continuing resolution for the first six months of FY 2013, sequestration compelled us to reduce our operations, including cancelling five ship deployments, delaying deployment of the USS HARRY S TRUMAN strike group by six months, and creating a significant maintenance backlog that has carried over into FY 2014. The Navy was forced to defer required maintenance on 16 airframes and 55 engines/engine modules and reduce non-deployed flying hours, thereby jeopardizing planned aircraft modernization, mission system software capability improvements, fatigue-life management, depot support, and the flight-hour program which maintains pilot proficiency and readiness. While this preserved forward deployed activities and training for next-to-deploy forces, it increased risk in the readiness of our non-deployed units which will take years to recover.
>
> Sequestration also resulted in the furlough of DON civilian employees for six days, which when combined with a hiring freeze, reduced our maintenance and sustainment capacity by taking away logisticians, comptrollers, engineers, contracting officers, and planners. Navy facilities sustainment, restoration and modernization was reduced by approximately 30% and base operations, including port and airfield operations, was reduced by approximately 20%.[9]

September 18, 2013, Navy Testimony

At a September 18, 2013, hearing before the House Armed Services Committee on planning for sequestration in FY2014 and perspectives of the military services on the Strategic Choices and Management Review (SCMR), Admiral Jonathan Greenert, the Chief of Naval Operations (CNO) testified, in a statement similar to the one he used at the November 7, 2013, hearing discussed above, that

> the FY 2013 sequestration reductions compelled us to reduce our afloat and ashore operations and created a significant afloat and ashore maintenance backlog. However, the effects were barely manageable due to authorization to reprogram funds into appropriate maintenance accounts, and we were able to use prior-year investment balances to mitigate reductions to investment programs. Impact to Navy programs, caused by the combination of a continuing resolution and sequestration, included:
>
> - Cancelled five ship deployments.
> - Delayed deployment of [the aircraft carrier] USS HARRY S TRUMAN strike group by six months.
> - Planned inactivation, instead of repairing, [the attack submarine] USS MIAMI due to rising cost and inadequate maintenance funds.
> - Reduced facilities restoration and modernization by about 30%.
> - Furloughed DON [Department of the Navy] civilian employees for 6 days, which, combined with a hiring freeze, reduced our maintenance and sustainment capacity by taking away logisticians, comptrollers, engineers, contracting officers, and planners.
> - Reduced base operations, including port and airfield operations, by about 20%.
> - Cancelled the Blue Angels' season and most non-essential port visits for Fleet Weeks.[10]

September 5, 2013, CNO Remarks

At a September 5, 2013, event at the American Enterprise Institute (AEI), Admiral Jonathan Greenert, stated that

> ...we have 285 ships today, about an average of 95 [of those ships] deployed. This is about a 90 day average. ... But we're about 10 [ships] down in this deployment piece. I.e., we were about 105 deployed [ships] about a year ago, and that is a factor of this last year of the budget limitations that we've had....
>
> This rebalance to the Asia Pacific, despite operations around the world, despite Mideast operations, continues. It has slowed down. The continuing resolution of '13 and sequestration has slowed us down but it's moving ahead....
>
> The effect of sequestration, the continuing resolution in FY13 pretty much came about as we predicted and as we testified to. The budget reduction was about $11 billion to us, to the Navy. And we were fortunate enough to reach back to prior year money, which hadn't been fully obligated, and pull that forward into '13 and it helped mitigate that. That's a one-time operation that we were able to do.
>
> We in fact had to cancel five ship deployments in FY13. [And regarding] Our surge capacity, the ability to respond here [indicating an overseas location on a briefing slide]. [using] Those ships that are back here in the continental U.S. Usually we have three carrier strike groups and three amphibious ready groups able to respond within a week. We have one [of each] now. That's going to be the story in FY14 as we look ahead. So it's a reduction in surge [capacity]. That's where a lot of the, if you will, the reductions in the budget kind of manifested themselves.
>
> We've done very little shore maintenance upgrades. If there's an area that I'm concerned about and I have to watch closely, it is our shore readiness. This is where we're taking a lot of the reductions and we've got to be careful of that.[11]

August 1, 2013, Navy Testimony

At an August 1, 2013, hearing before the Seapower and Projection Forces subcommittee and the Readiness subcommittee of the House Armed Services Committee on Navy surface ship maintenance and readiness, the Navy testified:

> ***Current readiness***
>
> The combination of the continuing resolution and sequestration put twenty three FY13 surface ship availabilities at risk, and represented the most immediate threat to surface ship readiness. We were able to restore all but eight availabilities when the FY13 appropriations bill was passed, and we appreciate the support of Congress on a reprogramming which will fund the last eight availabilities.
>
> The FY13 appropriations bill with sequestration left the Navy with a $4.1 billion shortfall in our Operations and Maintenance (O&MN) accounts compared with the President's 2013 budget submission. This has had an impact on Fleet operations and readiness in FY13, and will carry over into FY14. Specifically, it has degraded our ability to provide the level of global presence and surge capacity that we have executed over the last several years.
>
> The decreased presence is apparent in our reduction of deployed carrier strike groups, as well as a reduction in Southern Command and European Command deployments. For example, of the ten vessels scheduled to conduct deployments to Southern Command this fiscal year, only three will complete their deployments as planned. We will continue to provide ready forces to execute the highest priority deployments, providing the Combatant Commanders with the presence and capabilities they need most to execute the Defense Strategic Guidance. However, reduction of presence or elimination of deployments to any region is noticed by both our allies and potential adversaries, degrading not only our ability to build and foster cooperative relationships with our maritime partners, but also reducing our capability to ensure operational access and freedom of action. You cannot surge trust; rather you have to be there, building it every day.
>
> The decrease in our surge capacity is less apparent than reduced presence, but it still causes great concern due to the impact on war plans and contingency operations. The net effect is that surging our remaining surge capacity will likely lead to gaps in future regularly scheduled presence operations. Due to fiscal constraints, the Navy has been forced to prioritize maintenance and training for those forces deploying in FY14. Thus, those forces deploying after FY14 will receive reduced maintenance and training, decreasing our ability to surge these forces in case of emergency. This shortfall in surge capacity will be problematic if our forces are required to respond to contingencies. Currently, our surge forces are restricted to the forces trained and equipped for the next deployment, while the rest of the Fleet is in a training and material readiness status below "ready to deploy in all warfare areas."[12]

July 19, 2013, Navy Press Briefing

At a July 19, 2013, press briefing, Admiral Jonathan Greenert, the Chief of Naval Operations, stated that

> So could I get the Navy slide today up there, please? Thank you. I think you have this as part of the handout, but this is where your Navy is today. Presence remains our mandate. This is what we're mostly about. And it's a central element of our defense strategic guidance.

As you can see, we have about 95 ships deployed, and about 3,700 operational aircraft are also out there. I'll tell you, since sequestration sort of set in with the impact of a continuing resolution, we're down about 10 ships from, say, about a year ago or actually several months ago, forward deployed. So there is an impact....

In the Southern Command, sequestration has effectively caused us to reduce our combatant ships to zero. You can see up there [on the slide]. I tell you, there are other naval forces in the region, non-combatant ships and other forces, but we're zero [ships] today [in the Southern Command area]. And that was a deliberate decision approved by the Secretary of Defense as part of our global force management operation.

I would tell you, it'll undulate a little bit somewhere around zero, one or two as we go through the process....

So a little bit on the—the budget and sequestration, it's on my mind daily, and the fiscal year '13 appropriation bill helped us quite a bit. We're out there today with one carrier strike group and one amphibious ready group deployed to each theater, so we've got one in the Arabian Gulf region and one in the Western Pacific of each. But the issue is, the backup— that would be the surge force—we're not where we need to be in that regard. We have today one carrier strike group and one amphibious ready group ready to deploy with all the capabilities that we have in our covenant to our combatant commanders. The rest of the fleet is not ready to deploy with all the capabilities that are needed that we would normally have in our fleet response plan, and that's really the issue that we have there.

A year ago, I would tell you, we had three carrier strike groups and three amphibious ready groups ready to surge. And if there were a contingency, we had to take on a large operation, the surge force would be a concern, and the concern would be the capabilities that we would bring and whether or not they were the right capabilities.

It may not be readily apparent to many, because as you look out there, you say, hey, it kind of looks the same out there. But it's—it's the surge issue, and it's a real issue.

Now, for the remainder of the year, this year, this fiscal year, we'll be adjusting our operations and maintenance spending to meet, really, the FY [fiscal year] '14 deployments. The kids are training and doing maintenance this year for next year's deployments. And we have a plan in place, working with the Congress on a reprogramming, to restore the maintenance availabilities that we had set aside due to our budgetary issues earlier in the year. And we're trying to maintain as much of our training operations as feasible, for the reasons I've said before.

My real concern tends to be the shore facilities and the shore readiness. They're taking the brunt of this reduction here in '13, both in the case of the continuing resolution and sequestration. And I'm looking very closely at this. We are effectively doing no projects, no restoration or modernization projects, and we have taken our base operations and our sustainment, the very basic things, down to really the minimum that we think is safe and appropriate.[13]

May 8, 2013, Navy Testimony

The Navy's prepared statement for the May 8, 2013, hearing on Navy shipbuilding programs before the Seapower subcommittee of the Senate Armed Services Committee stated the following regarding the impact of the March 1, 2013, sequester:

Sequestration ... reduced the Fiscal Year 2013 funding across all [Navy] accounts by roughly 8 percent, or about $10.7 billion total, thus directly impacting current and future readiness. The Navy is still reconciling the impact of this reduction; however, due to the mechanics of its implementation and the limits on Department-wide transfer authority authorized by the Fiscal Year 2013 Defense Appropriations Act, it is likely that the Department [of the Navy] will be compelled to reduce our near term forward presence, our planned depot maintenance and training to support future operational rotations, our procurement of ships, aircraft and weapons systems to meet our force structure and inventory requirements, and our investment in future capabilities and readiness; thus impacting our future readiness. Every major weapon system is impacted by sequestration in 2013 with impacts ranging from reducing quantities procured, delaying schedules (delivery and initial operational capability), deferring costs to future years (particularly in the case of executing programs, such as shipbuilding), and absorbing cost growth due to all of these impacts.[14]

During the discussion portion of the hearing, Sean Stackley, the Assistant Secretary of the Navy for Research, Development and Acquisition (i.e., the Navy's acquisition executive), made additional comments regarding the impact of the March 1, 2013, sequester in response to questions. Early in the discussion portion of the hearing, he stated:

Let me start by just describing 2013 since we're dealing with sequestration in the current budget year.

As I described at the outset, each of the program lines was impacted by sequestration. So in 2013, we are working line by line to mitigate the impact by either paying for the sequestration impact to prior year assets, which we had accumulated through the last four to five years of shipbuilding or trying to defer certain costs that we can defer to a later point in the cycle in order to keep the plan procurement on track. And there are certain cases where we're looking at do we need to de-scope certain items on the shipbuilding plan, but trying to keep the overall force structure number healthy.

In that approach, in fact, the [third] DDG 51 that was added by Congress in 2013 is, in fact, held up. Otherwise, the balance of the shipbuilding program is going forward admittedly at some risk, so some increased risk in terms of called budget executability. So we're trying to do this very mindfully if we allow sequestration to stop us in our tracks that will simply cause or cause the disruption to go through the roof.

So we're going to continue to work that destroyer with Congress so that you all understand its specific impact. We're going to continue to execute the balance of the shipbuilding program in '13.

We have brought forward the budget request in '14, which as you're well aware does not account for a sequestration in '14. And, in fact, in '14 and out in a more strategic review that's been accomplished under the direction and guidance of the secretary of Defense, we are looking at shipbuilding amidst all of the capabilities that the department is pursuing in terms of what are the impacts associated with reduction of the top line, and then what are the priorities that we need to bring forward in terms of funding those capabilities.

And central to all of that is driving down the cost of what we're procuring and driving out the cost of our doing business so that more of the dollars available can go towards capability.[15]

Later in the hearing, he stated that

across the board, sequestration affects everything that we do. So first, it has created a great deal of uncertainty in terms of our planning and allowing us to prioritize within a top line where our investments will go, so uncertainty creates an impact.

That uncertainty then trickles down into planning and procurement in the vendor base—first year [sic: tier] shipyards and then down in the vendor base below that. So we're having to keep an eye on ensuring that the vendor base that we're relying on in a longer term to support our shipbuilding requirements doesn't break as a result of delays or uncertainty from sequestration.

And then the most poignant impact is the dollar impact directly. Everything that we've been doing to try to reduce the cost of our shipbuilding program whether it's stabilizing requirements, whether it's trying to get stable production rates that allow investment by the shipbuilders, trying to wrap in a multiyear where we harvest the significant savings, putting that inside of a fixed price contract where we have confidence in the savings, sequestration unravels that to an extent.

So now what we have to do when we look at—prospectively at sequestration on the outyear budgets, we've got to fight for the priority that shipbuilding demands in order to hit the CNO's [Chief of Naval Operation's] requirement within the budget so that our efforts to reduce cost don't, in fact, go in a reverse direction as a result of sequestration and we end up with potential disruption, taking low shipbuilding rates that we have today and driving them lower and then ultimately driving those costs up. So we have to avoid that spiral that could occur if we unravel what we've been attempting to do with regards to stabilizing the shipbuilding plan over the last several years.[16]

Potential Impact of Possible Late 2013/Early 2014 Sequester on FY2014 Funding

Another potential issue for Congress concerns the potential impact on Navy shipbuilding programs of a possible sequester on FY2014 funding and unobligated prior-year funding) that might occur in late 2013 or early 2014 under the terms of the Budget Control Act of 2011 (S. 365/P.L. 112-25 of August 2, 2011).

November 7, 2013, Navy Testimony

At a November, 7, 2013, hearing before the Senate Armed Services Committee on the impact of sequestration on the national defense, Admiral Jonathan Greenert testified, in a statement similar to the one he used at the September 18, 2013, hearing discussed below, that

> Sequestration in FY 2014, particularly if combined with restrictions of a continuing resolution (CR), will reduce our readiness in the near-term and in the long-term exacerbate program impacts from budget reductions required under current law. The impacts below assume an approximate 10% cut to the Navy's budget; however, with military personnel accounts exempted, the cut could increase to 14% in all other appropriations. In addition, the restrictions imposed by a CR will reduce our ability to manage the impact of sequestration. The impacts of this reduced funding will be realized in two main categories of budget accounts: (1) operations and maintenance and (2) investments.
>
> (1) Operation and maintenance accounts will absorb a larger reduction than in FY 2013 from a smaller overall amount of money; in addition we must begin to address deferred "carry over" bills from FY 2013 that total approximately $2.3 billion over the next five years. Because we will prioritize meeting current presence requirements, we will be able to preserve 95% of the forward presence originally directed under the FY 2014 Global Force Management Allocation Plan (GFMAP). However, this is still only about half of the Combatant Commander's original request. To ensure adequate funding for the most important deployments, we were compelled to adjust the plan in advance of FY 2014 to remove the deployment of one CG to the Middle East, two salvage ships to Africa and South America and five large surface combatants to the Western Pacific. Most concerning, however, we will have two thirds less surge capacity in FY2014. Our

planned presence to meet the GFMAP in FY 2015 and beyond will also be at risk because maintenance cancelled in FY2014 may result in ships being unable to deploy in future years. At a minimum this lost maintenance will reduce the service life of these ships.

Because of the mechanics of sequestration, we cannot reprogram (move) funds from other accounts into operations and maintenance to make up for the sequestered amount. As a result, within operations and maintenance, we have to "go where the money is" and find savings in training, maintenance, civilian personnel, and shore facilities. The reductions in fleet training we are compelled to make will result in only one non-deployed CSG and one ARG trained and ready for surge operations – notionally without these reductions there would be three of each ready to deploy within about two weeks.

We will be compelled to cancel or defer planned FY 2014 fleet maintenance, including 34 of 55 surface ship maintenance periods totaling about $950 million – all in private shipyards – and 191 of about 700 aircraft depot maintenance actions. This missed maintenance will inevitably take time off the expected service life of our ships and aircraft, which in turn will make it harder to sustain even the smaller fleet we will have if the BCA caps remain in place for the long term. For example, a recent Center for Naval Analysis study estimated cancelling and not making up one maintenance period at the ten-year point in a DDG's life will shorten its overall service life by about five years.

We will be compelled to keep in place our freeze on hiring for most civilian positions. Ashore we will continue to conduct only safety-essential renovation and modernization of facilities, further increasing the large backlog in that area.

(2) Investment accounts will be particularly impacted by sequestration in FY 2014, and we will not be able to use prior-year funds to mitigate shortfalls as we did in FY 2013. Without Congressional action or mitigating circumstances, the reductions imposed by sequestration and the limitations of a CR will compel us to:

- Cancel planned FY 2014 procurement of an SSN, an LCS and an AFSB; also, delay an SSN planned for FY 2015 procurement. Each of these would further worsen the reduction in fleet size, described earlier in this statement, that the BCA would compel us to make over the long term.
- Delay the planned start of construction on the first SSBN(X) from FY 2021 to FY 2022. This would cause us to be unable to meet U.S. Strategic Command presence requirements when the Ohio-class SSBN retires.
- Cancel procurement of 11 tactical aircraft (4 EA-18G Growler, 1 F-35C Lightning II, 1 E- 2D Advanced Hawkeye, 2 P-8A Poseidon, 3 MH-60 Seahawk) and about 400 weapons, exacerbating future BCA-driven reductions in our capabilities to project power despite A2/AD threats.
- Delay delivery of USS GERALD R. FORD (CVN-78) by two years, extending the period of 10 CVN in service, and lowering surge capacity.
- Delay the mid-life overhaul of USS GEORGE WASHINGTON (CVN 73) scheduled for FY 2016, disrupting today's "heel-to-toe" CVN overhaul schedule and reducing near-term CVN capacity.

In order to avoid or remedy some of the FY 2014 impacts described above, we need Congress to approve authorization and appropriations bills. This would enable the Navy to transfer funds, pursue innovative acquisition approaches, start new projects, increase production quantities, and complete ships. This would:

- Keep SSBN(X) on schedule to sustain required SSBN capacity after the Ohio class begins to retire.

- Buy two Virginia class SSN in FY 2014 as planned and keep FY 2015 SSN procurement on schedule. These actions will help maintain our undersea dominance and ability to project power despite A2/AD threats.
- Protect CVN-73's mid-life overhaul and complete CVN-78 on time to sustain CVN capacity.
- Build the planned AFSB in FY 2014, which is needed to meet DSG and combatant commander presence requirements for CT/IW capability.
- Restore half of the cancelled surface ship maintenance availabilities to protect FY 2015 presence.[17]

October 23, 2013, Navy Testimony

At an October 23, 2013, hearing before the Tactical Air and Land Forces subcommittee of the House Armed Services Committee on the impacts of a continuing resolution (CR) and sequestration on acquisition and modernization, Department of the Navy officials testified that

> the effect of another continuing resolution and sequestration-level reductions in FY 2014 would compound the impacts of the FY 2013 CR/sequestration. Operating under a CR in FY 2014, with corresponding restrictions on program funding and execution, significantly impairs our ability to effectively allocate resources and meet mission requirements. Compounding a CR with reductions to the FY 2014 budget in accordance with the mechanics of sequestration would remove from the Department its ability to provide warfighting capability and capacity with the measures of efficiency essential to balancing the requirements of the Defense Strategic Guidance with the fiscal constraints of the Budget Control Act (BCA) of 2011.
>
> In the near-term, the Navy and Marine Corps will prioritize preserving the global presence requirements set forth in the FY 2014 Global Force Management Allocation Plan (GFMAP). However, under sequestration-level funding, we will have less surge capability and our planned presence for FY 2015 and beyond will be at risk. Sequestration will compel us to forfeit long-term priorities to fund near-term readiness; resourcing training and maintenance of the next to deploy at the expense of those who will follow. Meanwhile, across the board reductions to investment accounts will slow production on factory floors across the defense industrial base adding cost and schedule to today's weapon systems; and equally critical, these reductions will drive delay into the development of those leading edge weapon systems that provide our warfighters with the asymmetric advantage they hold over our adversaries....
>
> The effects of an FY2014 CR and sequestration take away from the Department its ability to provide warfighting capability and capacity with the measures of efficiency essential to balancing the requirements of the Defense Strategic Guidance with the fiscal constraints under current law. Without Congress acting to change the current path, our warfighters will have less surge capability and our long-term priorities will be traded off to fund near-term readiness. Further, weapon system development timelines will be extended and costs will be higher, production unit costs will increase, and the risk to the long-term viability of the defense industrial base will increase.
>
> Our appeal is that Congress complete its work on the FY 2014 defense authorization and appropriations bills and eliminate sequestration before we are driven to irreversible actions which impair our collective responsibility to provide for the nation's defense.
>
> We understand the importance of resolving our fiscal challenges to ensure our nation's security and future prosperity and look forward to working with Congress to ensure our Navy and Marine Corps remain the world's preeminent maritime and expeditionary force.[18]

Regarding naval aviation programs in particular, the Department of the Navy officials testified at the hearing that

> Our FY 2014 Naval Aviation Budget request prioritizes several central themes: 5th generation fighter/attack capability; persistent multi-role intelligence, surveillance, and reconnaissance; supporting capabilities of electronic attack, maritime patrol, and vertical lift; robust strike weapons programs; and targeted modernization of the force for relevance and sustainability. It enables Naval Aviation to continue recapitalization of our aging fleets of airborne early warning, maritime patrol, electronic attack, and vertical lift platforms.
>
> FY 2014 sequestration is estimated to impose approximately a 10% cut to the DONs top-line budget. With military personnel exempted by the President, the reduction increases approximately to 14% in all other appropriations. With this loss of obligational authority, no use of prior year investment funds to mitigate shortfalls, and a need to address deferred FY 2013 carry over bills, all aspects of Naval Aviation will be negatively impacted, including both current readiness and development of future capabilities, eroding our margin of military superiority.
>
> A FY 2014 CR further creates several impacts, becoming more serious the longer the CR continues into the fiscal year. Under a CR, we are unable to use funds for new procurements or increase production rates above those sustained in FY 2013. Major programs that will be affected include: F-35 Joint Strike Fighter (JSF), MQ-8 FireScout, EA-18G Growler, Joint Precision Approach and Landing System (JPALS), and Advanced Precision Kill Weapon System (APKWS II) integration with MH-60R Seahawk. The CR also prohibits new multi-year procurement contracts planned for the C-130J and E-2D aircraft and associated mission equipment, delaying awards to industry and impacting production lines, delivery schedules, and most importantly, additional cost levied on pressurized budgets.
>
> *Operation & Support (O&S) Impacts:*
>
> Maintenance is critical to ensure our aircraft meet their expected service lives, keep strike fighter inventories at required readiness, and to preclude a strike fighter inventory shortfall above manageable levels. Our FY 2014 President's Budget submission included the resources necessary to continue Service Life Extension Program (SLEP) efforts and modifications on legacy F/A-18A-D Hornets to extend their service lives from 6,000 flight hours to 10,000 flight hours. Sequestration, however, will require us to cancel or defer approximately 200 of 700 aircraft and 580 engine planned depot maintenance events; eroding the service life of the aircraft and making it more difficult to sustain even the smaller fleet we will have if the reduced discretionary funding caps remain in place. If the current law levels remain long term, the maintenance backlog will continue to compound, eventually leaving Navy with insufficient aircraft available in inventory to meet deployment and training readiness objectives.
>
> Marine Corps aviation readiness will degrade with reductions to the flying hour program and depot maintenance in 2014. In 2013, approximately 70% of Marine squadrons met minimum deployable combat readiness. By 2015, we project this number will be reduced to approximately 60% as a result of sequestration.
>
> Marine F/A-18A-D Hornet squadrons face the biggest challenge with regards to funding reductions and depot maintenance backlogs. In September 2013, approximately half of the available F/A-18 inventory was in an "out of reporting" status. This results in non-deployed squadrons having only seven aircraft available for tasking when the squadron requirement is 12 aircraft. It is estimated that by January 2015 that number will be reduced to four aircraft available per non-deployed F/A-18A-D squadron.

Personnel Impacts:

The DON relies on a 'total force' of military (active and reserve) and civilian personnel to execute its mission. Even with the exemption of military personnel accounts from sequestration in FY 2014, the DON will continue to drawdown military end strength as noted in the FY 2014 President's Budget Request. However, if the reduced discretionary caps continue in force past 2014, we will not be able to afford our planned force structure, putting the nation's ability to respond to crisis around the globe at risk.

The impacts to our civilian workforce will affect every state and be detrimental to the employees, their families, the DON mission, and local economies. While we will make every effort to protect civilians from another furlough in FY 2014, sequestration-level funding forces us to consider all civilian force shaping tools, to include Voluntary Separation Incentive Pay, Voluntary Early Retirement Authority, Reductions in Force (RIF), continued hiring restrictions, and reductions or eliminations of bonuses. With the direct loss of labor- hours, the workforce impacts also lead to inefficiencies caused by loss of learning, productivity losses, cost increases driven by lengthening schedules, increased burdens on military personnel, and lower morale – all of which translate to reduced readiness.

Investment Account Impacts:

Due to the mechanics of its implementation and the limits on Department-wide transfer authority, sequestration will again impact every program and system, forcing reductions in procurement quantities, delays in schedules (delivery and initial operational capability), deferral of costs to future years, and unnecessary cost growth.

The DON remains firmly committed to the JSF program as an essential platform in our immediate and long-range Navy and Marine Corps aviation strategy and the nation's security. However, these funding constraints would compel us to reduce aircraft procurement by one F-35B Short Take-Off and Vertical Landing variant and one F-35C Carrier Variant and delay JSF Block IIIF development, test, and evaluation flights resulting in increased risk of meeting planned initial operating capability dates. As we carefully monitor strike fighter inventory requirements and projected availability, these reductions and delays also add increased risk to our ability to meet operational demands for expeditionary strike and maintain a complementary mix of strike fighter aircraft.

Sequestration will compel us to reduce E/A-18G Growler procurement by up to four aircraft in FY 2014, relative to the budget request, which will decrease attrition reserve aircraft set aside for the Navy's expeditionary force. Although this reduction will not delay the transition from EA-6B Prowlers to EA-18G Growlers, it will increase the risk associated with inventory requirements to offset loss of aircraft over the life of the EA-18G. For more than half a century, the DON has been the leader in Airborne Electronic Attack (AEA) and this asymmetric naval capability remains in high demand by the joint force. This reduction would limit our only tactical AEA capability and reduce our ability to rapidly respond to emergent operations on short notice. With FY 2014 as the last year of production, there will be no opportunity to adjust in the future.

E-2D Advanced Hawkeye procurement will be reduced by one aircraft relative to the request in FY 2014. If the current law caps remain long-term, Navy would continue to field the Navy Integrated Fire Control – Counter Air (NIFC-CA) network with E-2D in 2015, but may have to reduce the number of CVWs that have this capability in 2020 from six in our FY 2014 President's Budget submission to four. The completion of transition to the E-2D would delay by three years from 2023 to 2026.

P-8A Poseidon procurement may be reduced by up to two aircraft in FY 2014. If the current law caps remain long-term, one potential scenario would not allow our development of capabilities to project power to stay ahead of potential adversaries' Anti-Access/Area Denial (A2/AD) capabilities, as our undersea capabilities will be slowed. For example, attainment of the required P-8A inventory (117) will be delayed from 2019 to 2020, and the completion of the transition from the P-3C to the P-8A will be delayed from 2019 to 2020. This would also increase the sustainment cost of supporting legacy P-3C aircraft.

MV-22B Osprey procurement will be reduced by up to three aircraft if sequestered in FY 2014. This action would threaten the V-22 Multi-Year Procurement (MYP) contract and require price renegotiation for all remaining aircraft. This significantly reduces the estimated $1 billion in cost savings over the span of the current MYP contract. The increase in unit costs will further reduce the quantity of aircraft that the Marine Corps can afford to purchase, delaying full stand-up of future MV-22 squadrons, and Full Operational Capability.

Sequestration-level funding will also force reductions and delays in Unmanned Aerial Systems (UAS) programs. The Marine Corps will be required to reduce RQ-21 Small Tactical Unmanned Air Systems (STUAS) procurement by one system, resulting in a production rate below its minimum sustainment rate and increased system cost. This will put the program at risk, eliminating a critical capability for persistent ship and land-based ISR support for tactical-level maneuver decisions and unit level force defense and force protection missions. The MQ-4C Triton UAS (formerly known as BAMS for Broad Area Maritime Surveillance) initial operational test and evaluation will be delayed up to 12 months, delaying the persistent maritime intelligence, surveillance, and reconnaissance coverage it will provide to increase our maritime domain awareness and presence.

Overall, relative to the Budget request, sequestration in FY 2014 would result in the loss of an estimated 25 aircraft (2 F-35, 4 EA-18G, 3 MV-22, 4 H-1, 1 E-2D, 2 P-8A, 1 KC-130J, 3 MH-60, 1 UC-12, and 4 JPATS) across the Navy and Marine Corps and the loss of multi- year procurement contract savings. Critical development and delivery of capabilities will be delayed, putting our warfighters in increased risk against technologically advanced adversaries, especially over the long-term.

Sequestration delays fielding critical capabilities, breaks multi-year procurements, and reduces the quantity of strike and ship self-defense weapons. A broad spectrum of key strike weapons would be reduced, to include as many as seventy-three Joint Standoff Weapons; forty-six Tomahawk cruise missiles; fifty-one Hellfire weapons; forty-two Advanced Anti- Radiation Guided Missiles; and thirty-four AIM-9X Sidewinder missiles; thereby reducing our overall strike capability and capacity. We would also be required to make cuts in ammunition and training munitions that would be below minimum inventory requirements thereby placing Fleet training and readiness at further risk. These cuts would affect General Purpose Bombs, Practice Bombs, Air Expendable Countermeasures, Airborne Rockets, and Cartridge Actuated Devices/Propellant Actuated Devices.

If current law level caps continue, we would also see key components of the planned improved air-to-air infrared kill (IR) chain that circumvents adversary radar jamming, delayed by two years and the new, longer range AIM-9X/Block III missile delayed by up to two years. Improvements to the air-to-air radio frequency (RF) kill chain also would be slowed down as F/A-18E/F Block II Super Hornet anti-jamming upgrades would be delayed to 2020 and the equipping of all Pacific carrier air wings with the medium-range AIM-120D missile delayed by two years to 2022.[19]

September 18, 2013, Navy Testimony

At a September 18, 2013, hearing before the House Armed Services Committee on planning for sequestration in FY2014 and perspectives of the military services on the

Strategic Choices and Management Review (SCMR), Admiral Jonathan Greenert testified, in a statement similar to the one he used at the November 7, 2013, hearing discussed above, that

> Sequestration in FY 2014, particularly if combined with restrictions of a continuing resolution (CR), will reduce our readiness in the near-term and exacerbate program impacts from budget reductions required under current law in the long-term. The impacts below assume an approximate 10% cut to the Navy's budget; however, with military personnel accounts exempted, the cut could increase to 14% in all other appropriations. In addition, the restrictions imposed by a CR will reduce our ability to manage the impact of sequestration. The impacts of this reduced funding will be realized in two main categories of budget accounts: (1) operations and maintenance and (2) investments.
>
> (1) Operation and maintenance accounts, if sequestered under a CR, will absorb a larger reduction than in FY 2013 from a smaller amount; in addition we must begin to address deferred "carry over" bills from FY 2013 that total approximately $2.3 billion over the next five years. Because we will prioritize meeting current presence requirements, we will be able to preserve 95% of the forward presence originally directed under the FY 2014 Global Force Management Allocation Plan (GFMAP). However, this is only about half of the Combatant Commander's original request. To ensure adequate funding for the most important deployments, we were compelled to adjust the plan in advance of FY2014 to remove the deployment of one CG to the Middle East, two salvage ships to Africa and South America and five large surface combatants to the Western Pacific. Most concerning, however, we will have two thirds less surge capacity in FY2014. Our planned presence to meet the GFMAP in FY 2015 and beyond will also be at risk because maintenance cancelled in FY2014 may result in ships being unable to deploy in future years. At a minimum this lost maintenance will reduce the service life of these ships.
>
> Because of the mechanics of sequestration, we cannot reprogram (move) funds from other accounts into operations and maintenance to make up for the sequestered amount. As a result, within operations and maintenance, we have to "go where the money is" and find savings in training, maintenance, civilian personnel, and shore facilities. The reductions in fleet training we are compelled to make will result in only one non-deployed CSG and one ARG trained and ready for surge operations – notionally without these reductions there would be three of each ready to deploy within about two weeks.
>
> We will be compelled to cancel or defer planned FY 2014 fleet maintenance, including 34 of 55 surface ship maintenance periods totaling about $950 million – all in private shipyards – and 191 of about 700 aircraft depot maintenance actions. This missed maintenance will inevitably take time off the expected service life of our ships and aircraft, which in turn will make it harder to sustain even the smaller fleet we will have if the BCA caps remain in place for the long term. For example, a recent Center for Naval Analysis study estimated cancelling and not making up one maintenance period at the ten-year point in a DDG's life will shorten its overall service life by about five years.
>
> We will be compelled to keep in place our freeze on hiring for most civilian positions. Ashore we will continue to conduct only safety-essential renovation and modernization of facilities, further increasing the large backlog in that area.
>
> (2) Investment accounts will be particularly impacted by sequestration in FY 2014, and we will not be able to use prior-year funds to mitigate shortfalls as we did in FY 2013. Without Congressional action or mitigating circumstances, the reductions imposed by sequestration and the limitations of a CR will compel us to:
>
> - Cancel planned FY 2014 procurement of an SSN, an LCS and an AFSB; also, delay an SSN planned for FY 2015 procurement. Each of these would further worsen the

reduction in fleet size, described earlier in this statement, that the BCA would compel us to make over the long term.
- Delay the planned start of construction on the first SSBN(X) from FY 2021 to FY 2022. This would cause us to be unable to meet U.S. Strategic Command presence requirements when the Ohio-class SSBN retires.
- Cancel procurement of 11 tactical aircraft (4 EA-18G Growler, 1 F-35C Lightning II, 1 E- 2D Advanced Hawkeye, 2 P-8A Poseidon, 3 MH-60 Seahawk) and about 400 weapons, exacerbating future BCA-driven reductions in our capabilities to project power despite A2/AD threats.
- Delay delivery of USS GERALD R. FORD (CVN-78) by two years, extending the period of 10 CVN in service, and lowering surge capacity.
- Delay the mid-life overhaul of USS GEORGE WASHINGTON (CVN 73) scheduled for FY 2016, disrupting today's "heel-to-toe" CVN overhaul schedule and reducing near-term CVN capacity.

In order to avoid or remedy some of the FY 2014 impacts described above, we need Congress to approve authorization and appropriations bills. This would enable Navy to transfer funds, pursue innovative acquisition approaches, start new projects, increase production quantities, and complete ships. This would:

- Keep SSBN(X) on schedule to sustain required SSBN capacity after the Ohio class begins to retire.
- Buy two Virginia class SSN in FY 2014 as planned and keep FY 2015 SSN procurement on schedule. These actions will help maintain our undersea dominance and ability to project power despite A2/AD threats.
- Protect CVN-73's mid-life overhaul and complete CVN-78 on time to sustain CVN capacity.
- Build the planned AFSB in FY 2014, which is needed to meet DSG and combatant commander presence requirements for CT/IW capability.
- Restore half of the cancelled surface ship maintenance availabilities to protect FY 2015 presence.[20]

September 14, 2013, Navy Blog Post

In a September 14, 2013, blog post, Rear Admiral John Kirby, the Navy's Chief of Information, stated:

> Lots of people are still talking about our ships in the eastern Mediterranean. And well they should.
>
> The President made it clear the other night that he still favors a strike, even as he pursues a diplomatic solution to the crisis in Syria.
>
> So, those destroyers are still out there. And they are still ready. As Secretary Mabus said just this week, "I guarantee you that if we are called upon to strike, we will strike hard and we will strike fast. Presence is what we do. It is who we are."
>
> But for all the attention we've received, two things keep getting lost in the clutter. First, those ships didn't surge to the Med. They were already there. We've been rotating two or three ballistic missile shooters through that region for several years now.
>
> And second, the forward presence they represent is no given. Well, maybe in the immediate future it is. But it's sure not going to be a given if sequestration remains the law of the land.

Now, some folks might say we're exaggerating about that—that we're using the Syria situation as a way to plug a better budget.

Well, consider this. If sequestration hits us square in the face again next year—and I see nothing in the cards that says it won't—we'll be able to keep one carrier strike group and one amphibious ready group deployed to both the Pacific and the Central Command regions. And we'll be able to keep another of each ready to go if needed. But not a whole heckuva lot more than that.

And those destroyers? Yea, we'll still keep them over there in the eastern Med. It's an important mission. But the only way we will maintain that presence is by basing destroyers in Spain, which we will do starting next year.

In fact, if sequestration continues into the next fiscal year, a lot of things are going to get real hard.

Another round of sequestration will cut the Navy's budget by 10 percent below our FY14 budget submission. Worse, it will again apply the arbitrary, across-the-board cuts to our accounts. And since the President exempted military manpower from the cuts, sequestration essentially whacks a whopping 14 percent from every other account.

But this time, we won't have prior-year funds to ease the sting. We won't have that flexibility because resources from previous years' budgets have already been spent. There's no piggy bank to crack open.

Well, what about reprogramming authority? Wouldn't that help? Sure it would. We'd love to be able to move some money around. But even with reprogramming authority under an FY14 Appropriations Bill—which, by the way, we don't have yet—sequestration would cut our operations and maintenance account by $4.6 billion instead of $5.6 billion. That's the account we use to keep those ships out there and those Sailors fully trained.

A cut of this size to that account—without reprogramming authority—will delay more than half our ship maintenance availabilities next year and reduce our training to "just in time," meaning our Sailors won't be ready until just before they leave.

In fact, we'll have to shut down two airwings for three months each and limit four others to only the minimum level of flying, the "tactical hard deck."

Not only will we have fewer ships, subs and aircraft ready to go if needed, we'll also lose $4.5 billion next fiscal year from the accounts we use to buy new ones.

Sequestration could cost us a littoral combat ship, an afloat forward staging base/mobile landing platform, and up to 25 aircraft (Prowler, JSF, Osprey and others) needed for our future fleet. It will also delay a Virginia-class submarine.

And if we start the fiscal year with a continuing resolution, we will also delay the refueling of USS George Washington and won't be authorized to finish building the Gerald R. Ford.

We'll be forced to award contracts later than planned. And we'll no doubt miss out on multi- year procurement savings. That means the ships and planes we do buy are going to cost us more and come off the production lines and shipyards a little slower. And that means a smaller fleet.

A few weeks ago, Defense Secretary Hagel explained the Strategic Choices Management Review (SCMR), and the actions that could partially offset some of the damage sequestration in FY14 promises.

"The bold management reforms, compensation changes, and force structure reductions identified by the Strategic Choices and Management Review," he said, "can help reduce the damage that would be caused by the persistence of sequestration in fiscal year 2014. But they won't come close to avoiding it altogether. If these abrupt cuts remain, we risk fielding a force that over the next few years is unprepared due to a lack of training, maintenance, and the latest equipment."

So, SCMR options—including an eight or nine carrier fleet—is not out of the realm of the possible.

Of course, this isn't just about the carriers. If the numbers of strike groups shrink, so would the number of surface combatants, amphibious ships and air wings.

Indeed, if sequestration continues through this decade, the Navy will have something like 38 fewer ships to meet Combatant Commanders' needs—needs which are unlikely to decrease.

And it won't just be the Combatant Commanders who suffer. Just this month, we let contracts in Grand Rapids, MI; Cedar Rapids, IA; Columbia City, IN; Rolling Meadows, IL, and North Charleston, SC, and a host of other cities and small towns across America.

A smaller fleet means fewer jobs and lost revenue for small businesses and an unstable industrial base.

Just look around. The Navy is in demand. Heavy demand. From the Asia-Pacific to the eastern Mediterranean, we really are, as the CNO likes to say, "where it matters, when it matters."

But with the prospect of another sequestered year ahead, I wonder just how long we'll be able to keep that up.[21]

September 5, 2013, CNO Remarks

At a September 5, 2013, event at the American Enterprise Institute (AEI), Admiral Jonathan Greenert stated:

So as we look into [FY]14, we had about $11 billion reduction in [FY]'13. Well, it's $14 billion in [FY]'14. It's ten percent.

We again, as you probably have read, have exempted manpower [from the FY2014 sequester] and it's the right thing to do. That means instead of [a] ten percent [reduction] for [the remaining] appropriations, the [remaining] appropriations that receive a reduction, [the] non-manpower [appropriation,] gets a 14 percent reduction.

What's going to be the impact? Well, subject to any action and help from the Congress as we move ahead, we'll probably have to cancel about half of our surface ship [maintenance] availabilit[ies], so that's 34 [canceled surface ship maintenance availabilities]. We'll cancel a lot of aircraft [maintenance] availabilities]. [That's] About 190 [aircraft maintenance availabilities]. Last year we canceled about 90. So we're getting a backlog that is concerning in that regard. It will take a long time.

If we restore the budget after [FY]'14 and say okay. I'll tell you what. You've got full up operations and maintenance budget, it will take about five years to get that backlog in aircraft maintenance down.

Navy-wide, we'll reduce training for those who are not going to deploy. That gets back to that surge element that I mentioned before, reduction in that regard. So those that are not deploying in FY14 will have less training.

We'll have some air wings that will go to what we call tactical hard deck, which means they'll fly, the pilots will fly and the air crews will receive training at a level which is really just above what we're comfortable with for safety of flight, and it gets them at a point where when they get ready to deploy they can ramp up relatively quickly.

There's nothing magic about it. It's a statistical point. Not where I want to be. I will be pursuing to see if we can get more flying money to train our pilots above that level and get our kids to sea more often in FY14 as we look for reprogramming opportunity, the ability to move money.

So again [i.e., as in FY2013,] our surge capacity, I predict, will be about one-third of the norm as we look into [FY]'14.

Remember, sequestration is, and we're assuming it occurs in FY14, it takes reduction in every single account. So our shipbuilding reductions, which we were able to attenuate with prior money in [FY]'13, will take a hit in [FY]'14. I would see a loss of ta [sic: a] littoral combat ship there, again without help; an afloat forward staging base [AFSB], which is an important part of our future; and advance procurement for our Virginia Class submarine; and for [aircraft] carrier [refueling complex] overhaul [RCOH].

We might lose two more [ships from the FY2014 shipbuilding account, namely], we might lose a submarine procurement in [FY]'14, and a destroyer, again, if we are unable to reprogram, to move money into those shipbuilding accounts from other accounts.

These will be challenges, these will be issues we'll be working with the Congress as we move ahead.

In aircraft [procurement] we'll probably lose about 25 aircraft. If you say what kind? We'll say there isn't any that won't probably be lost. Some helos, the P-8s, to F-35s, they'll all be affected by this because it goes to every single account.

[The] CIVPERS [civilian personnel] hiring freeze will probably continue through that period. There's a great potential we'll have to do a RIF, a reduction in force, in our civilian work force. So we'll start a voluntary program probably immediately in [FY]'14, offering programs for voluntary retirement to help attenuate the need to have to do a reduction in force.

The key to all of this is this transfer, being able to transfer money to get reprogramming for us, in order to have a balanced approach. If I were to estimate what I think we need, we need about a billion dollars to get into the operations and maintenance account, and about a billion dollars to get into the procurement account so we can get that into shipbuilding which will be our number one priority in the Navy.[22]

August 1, 2013, Navy Testimony

At an August 1, 2013, hearing before the Seapower and Projection Forces subcommittee and the Readiness subcommittee of the House Armed Services Committee on Navy surface ship maintenance and readiness, the Navy testified:

Future Readiness

The biggest challenge to future surface ship readiness during these fiscally constrained times is finding the correct balance between funding the necessary maintenance, to provide ready forces now, and executing life cycle maintenance that ensures the long term viability of our ships. As the Navy learned in the report of the 2010 Fleet Review Panel, the impact of delaying maintenance is significant, since the cost and duration of deferred repairs rise exponentially. The end result will be ships being decommissioned before their expected service life (ESL) due to degraded material condition.

Today, we are prioritizing current readiness over future readiness; however, this is not sustainable over the long-term. If we choose to neglect life cycle maintenance, the material condition of our ships will continue to degrade to the point that they may be unable to deploy or conduct routine operations, culminating in decommissioning ships before their ESL. Even when the trend is reversed and more funding is made available for future readiness (operations, training, and maintenance); it will take a significant amount of time to restore our readiness to levels that support both typical presence as well as surge requirements.

Future readiness will also be at risk if we fail to maintain the necessary capabilities and capacity in our ship repair industrial base. Variations in workload can cause peaks and valleys in the skilled labor demand of our industrial base. We cannot afford to lose the skilled labor force we need to maintain our highly complex ships.

Today, our maintenance and modernization process, to include government oversight of the private sector work, is extremely challenged by sequestration and furloughs. There have been disruptions to basic waterfront services. Inspection of critical check points is stressed. Testing is being delayed, as is the ability to place work on contract and modify it as circumstances warrant. Under the furloughs, our Regional Maintenance Centers are operating at approximately 64% manning for Contract Management and Oversight (CMO). Similar shortages are occurring in first responder technical assistance positions. As a result, it is estimated that availabilities will experience increases in duration of 20 or more days. Delays and impacts have been observed onboard USS ROSS, USS MILIUS, USS COMSTOCK, USS LABOON and USS MITSCHER.

We are concerned that the Navy's budget challenges will be greatly exacerbated in FY14 and beyond. FY14 sequestration will result in a $14B [billion] budget shortfall in the Navy, which will have a significant impact on our operations and maintenance accounts and will derail the efforts the Navy has made to restore the material condition of our surface fleet. Our current estimate is that approximately 64% of the FY14 surface ship availabilities will be at risk in the event of an FY14 sequester (absent reprogramming). These availabilities are necessary to repair broken equipment and upgrade obsolete systems needed for deployment, and to ensure each ship reaches its ESL of thirty-five to forty years.[23]

July 19, 2013, Navy Press Briefing

At a July 19, 2013, press briefing, Admiral Jonathan Greenert, the Chief of Naval Operations, stated that

Turning quickly to '14, we're still working to understand the fiscal situation, if sequestration is the—really, the rule of the day, and looking at all of the programs and all of the appropriations. The simple math is, it'll cause each account to go down 10 percent. That's what sequestration is. It's an algorithm.

For the Navy—for the Department of Defense, it's $52 billion. For the U.S. Navy, it's about $14 billion. Now, if manpower were excluded, as we did in '13, and we're still deliberating on that, all the accounts would go down 14 percent, because when you exclude one, it all goes to the other accounts.

The difference in '13 and '14 is, in '13, in our investments accounts, when sequestration set in, where we had to do the reductions, we had what we call prior-year money. We had money available from previous appropriations and laws coming in that we used to get us through '13. That's not available in '14. And so the impact on the investments accounts will be deeper cuts and a concern.

My goal, the secretary's goal would be to preserve shipbuilding and those ship contracts and those aviation contracts as much as possible, meet our forward presence requirements, as I mentioned before, and—because—and make sure we hold onto multiyear procurement. That is the most efficient way to purchase equipment and platforms. But, again, as we do that, that money—those reductions are real, and they'll take place in other accounts, as the case may be.

In fiscal year '14, as we look at it now, I think there will be a significant reduction in our surface ship [maintenance] availabilities,[24] subject to a change. About half of the 60 availabilities that we have planned will have to be deferred. You can't buy those back right away. [A] Ship is available for a maintenance period. If you don't do that maintenance period and that ship is then called out to do other things, it's a missed opportunity. So it takes time to get back.

Reduced certification training in FY '14 would affect FY '15 deployers, so we've got to reconcile that. We've got to watch our air wing readiness, folks coming back and shutting down due to inability to have money because they're not deploying right away. You can't get down far, deep into the readiness bathtub, as we would call it. It makes it that much harder to get out, so we'll have to watch that very closely. And, again, a reduced ship procurement would—and a break in a multiyear would cause each follow-on year's procurement of those ships that much more costly. So it's a very inefficient way to approach it.

The results for presence and for both '14 and '15, to kind of summarize, one carrier strike group, one amphibious ready group, each in each theater, Western Pacific and the Arabian Gulf. Our surge will be limited, really, to those that are next to deploy. They'll be ready to deploy with all the capabilities. But the rest of the fleet, regrettably, won't have the capabilities that we would notionally have and that we like to have in our plans to support.

The bottom line for all of this, I would ask you to think about, is the importance of being able to reprogram and being able to balance our accounts. Everything I kind of pass to you is all about an algorithm, a reduction across each account, and we need the ability, if you will, to be able to balance that.

Let me shift to people real quickly and then we'll go to questions and answers. As I mentioned before, they're the foundation of what we're about and what makes your Navy the finest Navy in the world. But we also have civilian shipmates, and regrettably, we're in the— we're enduring furloughs, and it's an impact. I felt it last week; I feel it this week. Just here at the staff, we feel it all around the Navy here or there.

Civilians are an integral part of our team. And we're committed to finding offsets to limit those effects whenever possible. We're looking at solutions, but it's got to be a Department of Defense-wide solution as we approach this.

We realize there's a financial burden. And it's regrettable. And it hurts our readiness, and it hurts our productivity, as well. And if sequestration continues into '14, we're going to need very closely—look very closely at that, at furloughs and, in my view, attempt to avoid it as much as feasible.[25]

During the subsequent question-and-answer portion of the briefing, the following exchange occurred:

Q: ... hello, sir. Two sequester-related questions, our favorite topic. One is, you were talking about capabilities we are missing, if we need to, you know, surge ships. Capabilities is a little vague. I presume it means more than, oh, gee, they're not sticker stocked in the onboard vending machines. It's probably a little bit more complicated and serious than that.

And, two, with this reprogramming you have pending, the Air Force got a bunch of grounded squadrons back in the air. Are you guys hoping, planning to have any similar reversals of bad fortune and get things back? You mentioned the ship's maintenance availabilities. And what capabilities have we lost? What can we get back?

ADM. GREENERT: When I speak to capabilities, Sydney, as an example, if you take an air wing, they do air-to-air, they do air-to-ground, they do a whole series of missions that the pilots are qualified to do, typically measured in how many cockpit hours you have in the air, how much you fly.

And so what will happen is, when you are unable to fully fund it and you're called upon, you have to reconcile, what missions will these pilots, in the case of an air wing, be qualified to do and certified to, so that we can report to the theater, hey, you're going to get an all up round in this particular mission in this regard. Similarly, with a destroyer or with a cruiser, what are they certified to do?

So we would have to tailor more, something we don't like to do, because the world is dynamic and they get a pretty big vote in what kind of operations we have, so we will typically certify our people for a range of operations and missions and certifications.

Q: So it's a matter of being trained to—to standard on the full range of things that you want people to do?

ADM. GREENERT: That's correct, because as you know—and in the past, we've had that— a carrier strike group or any of our units can swing theater to theater. Very different—well, somewhat different series of operations in, say, the Arabian Gulf, the North Arabian Sea versus the Western Pacific versus the Eastern Mediterranean.

STAFF: Dan?

Q: If you could just kind of relate what you said about piracy to sequestration and resources and so on, how—does that mean that you're not—the U.S. Navy will not be able to respond to that piracy problem as it might have two years ago or a year ago because of sequestration? And a related question. Could you go over again how U.S. naval capabilities are being affected and will be affected in the gulf, given this—these budget cuts that are coming?

ADM. GREENERT: The—the piracy question, the mission of counter- piracy is—the skill sets involved in that are more inherent in what we do as a Navy who is used to expeditionary operations around the world. So I would tell you, that—excuse me—that is a skill set that we can quickly revive and that doesn't concern me as much. In other words, those folks that are in theater, even if called on short notice, I think—I'm pretty comfortable in their counter- piracy capability.

Now, there is a skill set that is called visit, board, search and seizure. And that's basically coming—pulling alongside, either through a small boat or dropping through a helicopter, and that's kind of getting high end in that regard. That's a little different. That takes some practice. But typical, counter-piracy reporting, maybe bringing in a law enforcement detachment, we can do that relatively short order.

Your second question, capabilities affected in the Arabian Gulf as we move ahead. Again, it would depend on the mission, and it would depend on the requirements of the Central Command. The broad range of missions that we take there, ballistic missile defense, maritime intercept, sea control, as you mentioned, counter-piracy, all the way up to surface- to-surface missiles, you know, launching, counter- mining, that broad view we have to look at and perhaps tailor by—by unit, because the units—because of the money that we receive being so hard to predict out ahead. It would be a little different for each unit, so we have to watch this very closely on who we would send, if called upon in a contingency.

But let me enforce that those that we send over now, those that we'll send over in '14, they will be trained for the full range of missions. I'm right now talking about those that would surge if called upon in a broader contingency. Did I answer your question yet?

Q. Yeah, and then just could you again, how do you see the carrier group presence evolving over the next year or two, given—with sequestration?

ADM. GREENERT: It will be one in the Western Pacific carrier strike group, one in the Arabian Gulf for—you know, 1-0 [1.0], as we call, it one-one in each theater.

Q: And that's sustainable, even with sequestration?

ADM. GREENERT: For—through '14, as we currently are aligned in our budget. When I get into the '14 budget and I get into the details, may have to alter that, but that's—that's what I see right now in my planning.

Q: Just a follow-up on that, does that mean that this on carrier strike group presence in the gulf in FY '14 is—does that mean that the Navy relinquishes its goal of 1.7 ratio in the region?

ADM. GREENERT: Well, the relinquishing would be, I guess, a good choice of words. What that—what we have to do is we came forward and said, "Here is our fiscal situation. Here is our forward presence situation aligned to that fiscal situation. We have to make a choice."

In a given year, we could send 1.7 over, but that money spent on operations this year would be invested there, not in training. It just wouldn't be enough. And therefore, you're mortgaging next year's—or you're foreclosing, I should say, next year's deployment,

> because those folks will not be trained, unless somehow you scramble and get money through the year, and that would be high risk.
>
> So these are the—these are the discussions we had. We had these discussions early this year on the subject of the Truman deployment. And the simple choice was, we can continue the 1.7 and send the Harry Truman or—but that would be problematic for next year's deployment—or we could hold that, and it was determined overall that the right decision was—and it's a deliberate process, Matthew. It's the global force management process, approved by the secretary of defense.[26]

Later in the question-and-answer portion of the briefing, the following exchange occurred:

> Q: Admiral. I wanted to ask on sequestration (inaudible) and you have no prior (inaudible) you have lots of ships under contract. Are you considering canceling contracts? Or how will you take (inaudible) cancel contracts?
>
> ADM. GREENERT: It's a good question. What we need to do and what we're doing is, you sit down and you look at the contract, and what have you contracted—specifically have you contracted to do? What's in each line item? And how much does that cost? Is it feasible, for example, to—when you build a ship, for example, you get the ship, you get documentation, you get some ancillary gear, and you get some support, and you actually get some outfitting spares to get you through the first number of years until industry, who you contract for can procure the spares into the future.
>
> So you say, all right, can we continue—can we keep this under contract at what might be 14 percent less than we had originally in the budget and then deal with as you're building the ship, deal with it maybe later? Is that feasible? And we will do that with each of the builders. That's a notional example. It's not precise, but it's notional.
>
> These are the things that we'll have to do. Some of the things we buy have that sort of flex into them, and some of them don't. And if they don't, we'll lose the unit. And if you lose the unit, all the other ones in the contract, that price will go up accordingly.
>
> Q: Do you think, if you have sequestration in 2014, the goal of having 300 ships by the end of the decade, is that pretty much the goal?
>
> ADM. GREENERT: I would be hard-pressed to say it is out. If we can retain the ships under contract in that multi-year that I mentioned before, where each one costs significantly less because you're buying them in bulk, if you will, or in larger quantities, if we can't, then those—those multi-years sort of unravel. The unit price goes up. You have less money, and this just starts spiraling down. If sequestration continues beyond '14, I'm pretty confident that it would be—we would not be able to meet those goals. That's too much at $50 billion a year.[27]

May 8, 2013, Navy Testimony

At a May 8, 2013, hearing on Navy shipbuilding programs before the Seapower subcommittee of the Senate Armed Services Committee, Sean Stackley, the Assistant Secretary of the Navy for Research, Development and Acquisition (i.e., the Navy's acquisition executive), stated the following in response to a question about the potential impact of a sequester on FY2014 funding:

> Let me first say that Secretary [of Defense Chuck] Hagel has, in fact, chartered a Strategic Choices and Management Review to go directly at that question.
>
> The Armed Services Committee has sent out a letter to the secretary asking for a more discrete response to the potentially $52 billion [DOD-wide] impact [of sequestration] in 2014, and that response is being addressed in the real-time.

Now, let me just talk about shipbuilding and make some assumptions. If the $52 billion impact [equates to] about 10 percent [of the DOD budget] and if that was prorated across all her lines than in shipbuilding [sic: all the line items in shipbuilding?], we will be looking at greater than $1 billion reduction to the 2014 [shipbuilding] request. That's assuming no ability to prioritize our investments. That will be applying sequestration in '14 just like it was in '13 line by line.

In 2013, shipbuilding took about a $1.7 billion reduction. We're able to manage that to a great extent to the prior year [sic]. We had assets that we are building in the prior year, so we were able to pay off about a third to 40 percent of sequestration by liquidating those assets.

Of the balance, $1 billion to $1.2 billion, some of that we are able to reduce our requirements. Some of that we're going to have to work within '13 and some of it effectively (inaudible) [sic: we have to shift?] into the outyears.

If you then do that again in 2014, effectively we have pulled all the margin out of the system in shipbuilding. Where we had margin, we have pulled it out. So now if we double-down sequestration in 2014, the margin is gone and now we're looking at direct impacts to the— our ship procurement rates.[28]

Potential Impact on Size of Navy of Reducing DOD Spending to Lower BCA Caps through FY2021

Another potential issue for Congress concerns the potential impact on the size of the Navy of reducing DOD spending (through sequestration or regular appropriations activity) in FY2013- FY2021 to levels at or near the lower caps established in the Budget Control Act of 2011, or BCA (S. 365/P.L. 112-25 of August 2, 2011). Navy officials state that a decision to reduce DOD's budget to such levels would eventually lead to a smaller Navy.

November 7, 2013, Navy Testimony

At a November, 7, 2013, hearing before the Senate Armed Services Committee on the impact of sequestration on the national defense, Admiral Jonathan Greenert testified, in a statement similar to the one he used at the September 18, 2013, hearing discussed below, that

> Consistent with what the Deputy Secretary of Defense told this committee in August, if fiscally constrained to the revised discretionary caps, over the long term (2013-2023), the Navy of 2020 would not be able to execute the missions described in the DSG. There are numerous ways to adjust Navy's portfolio of programs to meet the BCA revised discretionary caps. These are currently under deliberation within the department. As requested, the following provides perspective on the level and type of adjustments that will need to be made.
>
> Any scenario to address the fiscal constraints under current law must include sufficient readiness, capability and manpower to complement the force structure capacity of ships and aircraft. This balance would need to be maintained to ensure each unit will be effective, even if the overall fleet is not able to execute the DSG. There are, however, many ways to balance between force structure, readiness, capability and manpower.
>
> One potential fiscal and programmatic scenario would result in a "2020 Fleet" of about 255- 260 ships, about 30 less than today, and about 40 less than Navy's PB-14 submission. It would include 1-2 fewer CSG, and 1-2 fewer ARG than today. This 2020 fleet would not meet the DSG requirements for the mission to Provide a Stabilizing

Presence. As a result, Navy would be less able to reinforce deterrence, build alliances and partnerships and influence events abroad.

- Navy would not increase our global deployed presence, which would remain at about 95 ships in 2020. The lethality inherent in this presence, based on ship type deployed, would be less than today's 95-ship presence.
- Navy would not increase presence in the Asia-Pacific, which would stay at about 50 ships in 2020. This would largely negate the ship force structure portion of our plan to rebalance to the Asia Pacific region directed by the DSG.
- Navy would not "place a premium on U.S. military presence in—and in support of—partner nations" in the Middle East, since presence would decrease and, assuming we use the same ship deployment scheme in the future, there would be gaps in CSG presence totaling 2- 3 months each year.
- Navy would still "evolve our posture" in Europe by meeting our ballistic missile defense European Phased Adaptive Approach (EPAA) requirements with four BMD-capable DDG homeported in Rota, Spain and two land based sites in Romania and Poland. Additional presence would still be provided by forward operating JHSV, MLP, AFSB and some rotationally deployed combatants.
- Navy would still provide "innovative, low-cost, and small-footprint approaches" to security in Africa and South America by deploying, on average, one JHSV and one LCS continuously to both regions and maintaining an AFSB in AFRICOM's area of responsibility.

In order to sustain a balance of force structure (current and future), modernization and personnel within our portfolio, continued compliance with the BCA revised discretionary caps would compel us to reduce our investments in force structure and modernization, which would result in a "2020 Fleet" that would not meet DSG direction in the following mission areas:

Counter Terrorism and Irregular Warfare (CT/IW). We would not have the capacity to conduct widely distributed CT/IW missions, as defined in the DSG. There would be inadequate LCS available to allocate to this non-core Navy mission, in the amount defined by the FSA and concurred upon by Special Operations Command.

Deter and Defeat Aggression. We would not be able to conduct one large-scale operation and also counter aggression by an opportunistic aggressor in a second theater. In this scenario, the fleet would have 9-10 CVN/CSG and 9-10 LHA/D and ARG. We would be able to sustain about one non-deployed CSG and one non-deployed ARG fully certified and able to surge on required timelines. Together, our presence and surge forces would be sufficient to conduct all missions associated with only one large scale operation, as defined today. This overall force and associated readiness would, however, be sufficient to execute Navy elements of the DSG mission to Conduct Stability and Counterinsurgency Operations.

Project Power Despite Anti-Access/Area Denial (A2/AD) Challenges. Overall, in this scenario, development of our capabilities to project power would not stay ahead of potential adversaries' A2/AD capabilities. We will not meet the projected capability requirements to assure Joint access in a plausible operational scenario in 2020 due to shortfalls, in particular, in air and missile defense:

- Some undersea capabilities will be slowed:

- Attainment of the required P-8A inventory (117) would be delayed from 2019 to 2020, and transition from the P-3C to the P-8A would be delayed from 2019 to 2020.
- Anti-submarine warfare combat system upgrades for DDGs and MFTA installations would not be affected.
- The LCS ASW Mission Package would be delayed from 2016 to 2017.
- Upgraded sonobuoys and advanced torpedo procurement would still equip all of our helicopters, SSN, and P-8A in the Western Pacific by 2018.
- Virginia Payload Module (VPM) would still be fielded in 2027 to enable Virginia-class SSN to replace SSGN that begin retiring in 2026.
- The LCS mine warfare mission package would still field its first increment in 2015 and the second in 2019.
- Air and missile defense improvements would be slowed:
- SEWIP upgraded electromagnetic sensing and upgraded jamming and deception capabilities would both be delayed one year (to 2015 and 2018, respectively). Both of these upgrades are required to counter advances in adversary anti-ship cruise missiles.
- The new Air and Missile Defense Radar (AMDR) would be delivered on only four ships, as compared to seven under our PB-14 submission, between 2021 and 2024.
- The Evolved Sea Sparrow Missile (ESSM) Block II would still be fielded in 2020, with 80 missiles being delivered to deployed ships.
- The F-35C Lightning II, the carrier-based variant of the Joint Strike Fighter, would still field in 2019 and join our CVW forward homeported in the Western Pacific in 2020. Overall, the number of F-35 procured would decrease by about 30 aircraft in 2020.
- All components of the improved air-to-air IR "kill chain" that circumvents adversary radar jamming would be delayed by two years. The Infrared Search and Track (IRST) sensor system would field in 2018 and the improved longer-range IRST would not deliver until 2021. The new longer-range AIM-9X Block III missile would not be fielded until 2023.
- Improvements to the air-to-air RF "kill chain" would be slowed down as F/A-18E/F Block II Super Hornet anti-jamming upgrades would be delayed to 2020. The longer-range AIM- 120D missile would still field in 2014 but equipping of all Pacific carrier air wings would be delayed by two years to 2022.
- The Navy Integrated Fire Control – Counter Air (NIFC-CA) network would still initially field with the E-2D Advanced Hawkeye in 2015, but only four CVW (compared to six in our PB-14 submission) would have it by 2020. Transition to the E-2D would be delayed three years to 2025.

Operate Effectively in Space and Cyber Space. Plans to recruit, hire and train 976 additional cyber operators and form 40 computer operations teams by 2017 would not be impacted. This is a priority in any fiscal scenario. However, the BCA's reduced funding levels would delay replacement of our cyber systems and decrease our ability to defend our networks.

Maintain a Safe, Secure, and Effective Nuclear Deterrent. We would still be able to sustain today's ballistic missile submarine (SSBN) force. The SSBN(X) would still deliver in 2030 to replace retiring Ohio class SSBN while meeting requirements for SSBN presence and surge. This is the top priority program for the Navy.

Defend the Homeland and Provide Support to Civil Authorities. We would still meet the capacity requirements for these missions.

Counter Weapons of Mass Destruction. We would still meet the presence requirements for this mission.

Conduct Humanitarian, Disaster Relief, and Other Operations. We would still meet the presence requirements for this mission.

The extent of the fiscal changes in the BCA, when compared to current program and budget levels, would compel Navy to request relief from several program mandates and force structure capacity limits, in order to sustain and build a fleet with a balance of ship types. For example, mandated limits govern the size of the force, minimum funding for certain activities and facilities, and changes to the number of personnel at a base.[29]

September 18, 2013, Navy Testimony

At a September 18, 2013, hearing before the House Armed Services Committee on planning for sequestration in FY2014 and perspectives of the military services on the Strategic Choices and Management Review (SCMR), Admiral Jonathan Greenert testified, in a statement similar to the one he used at the November 7, 2013, hearing discussed above, that

> Consistent with what the Deputy Secretary of Defense told this committee last month, if fiscally constrained to the revised discretionary caps, over the long term (2013-2023), the Navy of 2020 would not be able to execute the missions described in the DSG. There are numerous means and alternatives to adjust Navy's portfolio of programs. These are currently under deliberation within the department. As requested, the following provides perspective on the level and type of adjustments that will need to be made.
>
> Any scenario to address the fiscal constraints under current law must include sufficient readiness, capability and manpower to complement the force structure capacity of ships and aircraft. This balance would need to be maintained to ensure each unit will be effective, even if the overall fleet is not able to execute the DSG. There are, however, many ways to balance between force structure, readiness, capability and manpower. One potential fiscal and programmatic scenario would result in a "2020 Fleet" of about 255-260 ships, about 30 less than today, and about 40 less than Navy's PB-14 submission. It would include 1-2 fewer CSG, and 1-2 fewer ARG than today. With regard to the DSG and presence, in this particular scenario the "2020 Fleet":
>
> - Would not increase our global deployed presence, which would remain at about 95 ships in 2020. The lethality inherent in this presence, based on ship type deployed, would be less than today's 95-ship presence.
> - Would not increase presence in the Asia-Pacific, which would stay at about 50 ships in 2020. This would largely negate the ship force structure portion of our plan to rebalance to the Asia Pacific region directed by the DSG.
> - Would not "place a premium on U.S. military presence in—and in support of—partner nations" in the Middle East, since presence would decrease and, assuming we use the same ship deployment scheme in the future, there would be gaps in CSG presence totaling 2-3 months each year.
> - Would still "evolve our posture" in Europe by meeting our ballistic missile defense European Phased Adaptive Approach (EPAA) requirements with four BMD-capable DDG homeported in Rota, Spain and two land based sites in Romania and Poland. Additional presence would still be provided by forward operating JHSV, MLP, AFSB and some rotationally deployed combatants.
> - Would still provide "innovative, low-cost, and small-footprint approaches" to security in Africa and South America by deploying, on average, one JHSV and one

LCS continuously to both regions and maintaining an AFSB in AFRICOM's area of responsibility.

In order to sustain a balance of force structure (current and future), modernization and personnel within our portfolio, continued compliance with the current law discretionary caps would compel us to reduce our investments (force structure and modernization), which would result in a "2020 Fleet" that would not meet DSG direction in the following mission areas:

Counter Terrorism and Irregular Warfare (CT/IW). We would not have the capacity to conduct widely distributed CT/IW missions, as defined in the DSG. There would be inadequate LCS available to allocate to this non-core Navy mission, in the amount defined by the FSA and concurred to by Special Operations Command.

Deter and Defeat Aggression. We would not be able to conduct one large-scale operation and also counter aggression by an opportunistic aggressor in a second theater. In this scenario, the fleet would have 9-10 CVN/CSG and 9-10 LHA/D and ARG. We would be able to sustain about one non-deployed CSG and one non-deployed ARG ready and able to surge on required timelines to meet all missions associated with one large scale operation, as defined today.

Project Power Despite Anti-Access/Area Denial (A2/AD) Challenges. Overall, in this scenario, development of our capabilities to project power would not stay ahead of potential adversaries' A2/AD capabilities:

- Some undersea capabilities will be slowed:
  - Attainment of the required P-8A inventory (117) would be delayed from 2019 to 2020, and transition from the P-3C to the P-8A would be delayed from 2019 to 2020.
  - Anti-submarine warfare combat system upgrades for DDGs and MFTA installations would not be affected.
  - The LCS ASW Mission Package would be delayed from 2016 to 2017.
  - Upgraded sonobuoys and advanced torpedo procurement would still equip all of our helicopters, SSN, and P-8A in the Western Pacific by 2018.
  - Virginia Payload Module (VPM) would still be fielded in 2027 to enable Virginia- class SSN to replace SSGN that begin retiring in 2026.
  - The LCS mine warfare mission package would still field its first increment in 2015 and the second in 2019.
- Air and missile defense improvements would be slowed:
  - SEWIP upgraded electromagnetic sensing and upgraded jamming and deception capabilities would both be delayed one year (to 2015 and 2018, respectively). Both of these upgrades are required to counter advances in adversary anti-ship cruise missiles.
  - The new Air and Missile Defense Radar (AMDR) would be delivered on only four ships, as compared to seven under our PB-14 submission, between 2021 and 2024.
  - The Evolved Sea Sparrow Missile (ESSM) Block II would still be fielded in 2020, with 80 missiles being delivered to deployed ships.
  - The F-35C Lightning II, the carrier-based variant of the Joint Strike Fighter, would still field in 2019 and join our CVW forward homeported in the Western

Pacific in 2020. Overall, the number of F-35 procured would decrease by about 30 aircraft in 2020.

- All components of the improved air-to-air IR "kill chain" that circumvents adversary radar jamming would be delayed by two years. The Infrared Search and Track (IRST) sensor system would field in 2018 and the improved longer-range IRST would not deliver until 2021. The new longer-range AIM-9X Block III missile would not be fielded until 2023.
- Improvements to the air-to-air RF "kill chain" would be slowed down as F/A-18E/F Block II Super Hornet anti-jamming upgrades would be delayed to 2020. The longer-range AIM-120D missile would still field in 2014 but equipping of all Pacific carrier air wings would be delayed by two years to 2022.
- The Navy Integrated Fire Control – Counter Air (NIFC-CA) network would still initially field with the E-2D Advanced Hawkeye in 2015, but only four CVW, compared to six in our PB-14 submission, will have it by 2020. Transition to the E-2D would be delayed three years to 2025.

Operate Effectively in Space and Cyber Space. Plans to recruit, hire and train 976 additional cyber operators and form 40 computer operations teams by 2017 would not be impacted. This is a priority in any fiscal scenario.

Maintain a Safe, Secure, and Effective Nuclear Deterrent. We would still be able to sustain today's ballistic missile submarine (SSBN) force. The SSBN(X) would still deliver in 2030 to replace retiring Ohio class SSBN while meeting requirements for SSBN presence and surge. This is the top priority program for the Navy.

Defend the Homeland and Provide Support to Civil Authorities. We would still meet the capacity requirements for these missions.

Conduct Humanitarian, Disaster Relief, and Other Operations. We would continue to be able to support some missions with 2 ARG and 9 JHSV present overseas.

The extent of the fiscal changes in the BCA, when compared to current program and budget levels, would compel Navy to request relief from several program mandates and force structure capacity limits, in order to sustain and build a fleet with a balance of ship types. For example, mandated limits govern the size of the force, minimum funding for certain activities and facilities, and changes to the number of personnel at a base.[30]

September 5, 2013, CNO Remarks

At a September 5, 2013, event at the American Enterprise Institute (AEI), Admiral Greenert stated:

If you look out beyond [FY]'14, if you look at [FY]'15 through say the rest of that ten year period that the Budget Control Act is expected, we're looking right now in the building, in the Pentagon, among the services, we're building an alternative look from FY15-[FY]23 to say okay, if this continues and we just, rather than doing this year by year we just look at, what would this mean? And the Strategic Choices Management Review [SCMR] was a snapshot of that. It took scenarios and said well, if we looked at it this way, this would be the impact. It helped provide the Secretary of Defense that understanding of what kind of scenarios might be out there.

So we don't have "the" scenario now, but we have scenarios that we kind of lay in there.

My approach to this, it's going to be to make sure we maintain a credible and modern sea-based strategic [nuclear] deterrent. That's my number one program, the SSBNX [Ohio replacement ballistic missile submarine program] and the current Ohio [class SSBN] program, along with the command and control features and along with the [service life] extension of the [SSBNs' D-5 submarine-launched ballistic] missile [SLBM].

Next, we need to maximize forward presence. Using the forward deployed naval force, that's the concept we have in Japan and the concept we've been offered in Rota[, Spain] to bring ships, bring the Sailors and the families over there. We get great leverage from that. But also to forward station ships [in other locations] also. So forward presence.

Three, we've got to have adequate readiness. The ships that are deployed have to be ready because it's all that much important with a smaller budget with a smaller Navy, that Navy which is out and about is absolutely ready.

We've got to make sure that our asymmetric capabilities continue to be developed. The undersea domain, electronic warfare in the electromagnetic spectrum, the electromagnetic rail gun, our laser technology which is coming along. As you probably know, we are deploying a laser gun this coming summer, if you will, in the summer of [FY]'14 I should say, to the Arabian Gulf and we're looking to bring along the electronic rail gun.

[Two other priorities are] Cyber and people. Remember, people are an asymmetric advantage, so we've got to do that right.

We will reduce force structure in this plan, but we have to do it while preserving the right capacity to at least do one MCO [Major Contingency Operation, i.e., a major regional conflict] as we look out into the future, and we will have to reduce procurement. There's no doubt about it. I've got to look at the industrial base and make sure that as much as possible we're doing this in a deliberate and planned Navy.

That's kind of the principles of this look ahead.

Let me give you a scenario. If you say okay, what might you look like? And a scenario could be, in the future, if you take 2020 and you say what was your plan? Some of you may have seen this and say well, the plan [without sequestration] was to get to 295 ships with 116 planned to be deployed. That's in the red here [indicating on a briefing slide]. You say [in the future] future, if we had to retire a number of ships [due to defense spending being reduced to lower levels], [there] might [instead] be a 250 ship Navy in 2020, and able to get about 96 ships deployed.

A couple of thoughts about this. This is really leveraging operating forward, as I like to—It's one of my tenets in there. It's using the forward deployed naval force. Forward stationing ships. It's bringing along ships like the littoral combat ship [LCS] to join high speed vessel [JHSV], the afloat forward staging base [AFSB], and putting them in key areas of the world where we can leverage their usage. You get sort of a picture like that.

There is not a lot of surge [capacity] here. This is a Navy which, you can do the math, 96 [deployed ships] divided by [a total fleet of] 257 [ships]. You see how much is forward and out and about.

The reduction of our manpower associated with such a future where we reduce forces is completely connected to our force structure. What I mean by that is, we man equipment. That's the principle in the Navy with regard to manpower where some of the other services, they equip their manning. It's all about where our people are. And we have it about right now. We're just about where I want to be on the number of people per unit. So it's about retiring units, if you will, if you want to get a lot of savings out there as opposed to reducing people.

There's also a compensation entitlement reform and an overhead reduction which is part and parcel to a future look, and we are studying that closely.

I'll close now and say look, preserving presence is the key. That's our mandate, to be where it matters and to be ready when it matters, and we'll continue that rebalance to the Asia Pacific. We'll be moving ships forward. That's kind of a key element as I look into the future. And we've got to remain ready forward so we can do the things, like I showed you in the little graphic today. [The goal is to] Be able to respond quickly to what the nation needs.

But throughout it all, Sailors and our civilian work force, our civilian sailors, are going to be the asymmetric advantage, are going to assure that we keep a force which is whole and not hollow.[31]

August 1, 2013, Navy Testimony

At an August 1, 2013, hearing before the Seapower and Projection Forces subcommittee and the Readiness subcommittee of the House Armed Services Committee on Navy surface ship maintenance and readiness, Rear Admiral Thomas Rowden, the Director for Surface Warfare, testified that

> Keeping our equipment properly operating and training sailors takes money and time. When required funding is limited through sequestration or for other reasons, we are forced to make hard decisions about what gets fixed and what training is completed.
>
> Make no mistake. We will deploy ready ships. But our ability to respond to contingencies and surge additional ships to a crisis could be reduced.
>
> In the long term, if we do not maintain our ships, the Navy will be forced to decommission ships before their expected service life—or before they reach their expected service life. We estimate that if sequestration continues over the long term, by 2020 we will be reduced from 295 ships down to 257 ships. Over the same period, we do not anticipate the combatant commanders' demand for ready forces will decrease.[32]

Later in the hearing, he stated: "There is no doubt, I think, that given where we are proposing to go with sequestration that it will reduce the number of ships that we have in the Navy, and I think probably by 2020, we're looking at about 257 ships."[33] Still later in the hearing, he stated that

> it goes back to the—the deferral of maintenance and the bow-waving of maintenance into successive years. If we have to defer X number this year and that pushes into the following fiscal year and we have to defer Y, eventually we get to the point where the cost to maintain those ships goes to the point where it's just not—it's not cost-effective to maintain those ships and therefore you start to remove them out of service early.
>
> Based on what we've seen, there's a potential that we could go down to as many as 257. I haven't estimated it beyond that. I don't know whether Admiral Matthews[34] has anything beyond the 2020 timeframe. But looking at where we think we may have to go in the next seven years, we're looking at about—we think we're looking at about 257 ships by the time we get to 2020.[35]

July 31, 2013, DOD Press Briefing

At a July 31, 2013, press briefing on the Strategic Choices and Management Review (SCMR)—a DOD study of potential options for accommodating reductions in future defense funding levels— Secretary of Defense Chuck Hagel stated that "sequester-level cuts" to defense funding could, among other things, "reduce the number of carrier strike groups from 11 to 8 or 9.... "[36]

April 16, 2013, Navy Testimony

At an April 16, 2013, hearing before the House Armed Services Committee on the Department of the Navy's proposed FY2014 budget, Admiral Greenert, in answer to a question about the impact of sequestration on the Navy, stated:

> Mr. Chairman, as I—as I look at the numbers, and I think you're—you're talking, assuming a sequestration, $500 billion. First thing I do is, and most important, we provide forward presence. And it's—I can't provide—I cannot meet the—the current Global Force Management Allocation Plan with those numbers. So I don't know what number I would be at. It would be on the order 30 ships [less than today], you know, as I look at a balanced reduction in that regard, less than the number of ships that I have today.
>
> So, let's say 250 ships if I'm at 280 today.[37]

February 12, 2013, Navy Testimony

At a February 12, 2013, hearing before the Senate Armed Services Committee on the impacts of sequestration and/or a full-year continuing resolution on the Department of Defense, the Navy testified that:

> In addition to sequestration for FY13, the BCA also required the lowering of the discretionary caps for FY14 through FY21. Beyond FY13, if the discretionary cap reductions are sustained for the full nine years, we would fundamentally change the Navy as currently organized, trained and equipped. As time allows, we will take a deliberate and comprehensive approach to this reduction, based on a reevaluation of the Defense Strategic Guidance. In doing so, I will endeavor to: (1) ensure our people are properly resourced; (2) protect sufficient current readiness and warfighting capability; (3) sustain some ability to operate forward by continuing to forward base forces in Japan, Spain, Singapore and Bahrain, and by using rotational crews; and (4) maintain appropriate research and development.
>
> As I indicated last year to the Senate Armed Services Committee (SASC), under a set of fiscal circumstances in sequestration, our Navy may be a fleet of around 230 ships. That would be a loss of more than 50 ships, including the loss of at least two carrier strike groups. We would be compelled to retire ships early and reduce procurement of new ships and aircraft. This would result in a requisite reduction in our end strength. Every program will be affected and as Secretary Panetta noted in his 2011 letter to Senators McCain and Graham, programs such as the F-35 Lightning II, next generation ballistic missile submarine and Littoral Combat Ship might be reduced or terminated.[38]

Appropriate Future Size and Structure of Navy in Light of Strategic and Budgetary Changes

Another potential oversight issue for Congress concerns the appropriate future size and structure of the Navy. Changes in strategic and budgetary circumstances have led to a broad debate over the future size and structure of the military, including the Navy. Changes in strategic circumstances include, among other things, the end of U.S. combat operations in Iraq, the winding down of U.S. combat operations in Afghanistan, and China's military (including naval) modernization effort.[39] Changes in budgetary circumstances center on reductions in planned levels of defense spending resulting from the Budget Control Act of 2011, or BCA (S. 365/P.L. 112-25 of August 2, 2011).

On January 5, 2012, the Administration announced that, in light of the end of U.S. combat operations in Iraq, the winding down of such operations in Afghanistan, and developments in the Asia-Pacific region, U.S. defense strategy in coming years will include a stronger focus on the Asia-Pacific region.[40] Since the Asia-Pacific region is primarily a maritime and aerospace theater for the DOD, this shift in strategic focus is expected by many observers to result in a shift in the allocation of DOD resources toward the Navy and Air Force. DOD officials have indicated that efforts to support a stronger focus on the Asia-Pacific region will be protected if planned levels of DOD spending in future years are reduced as a result of the BCA or other legislative action.

Table 7. Recent Study Group Proposals for Navy Ship Force Structure

<table>
<tr><th>Ship type</th><th>Navy's 306-ship goal of January 2013</th><th>Project on Defense Alternatives (PDA) (November 2012)</th><th>Heritage Foundation (April 2011)</th><th>Cato Institute (September 2010)[a]</th><th>Independent Panel Assessment of 2010 QDR (July 2010)</th><th>Sustainable Defense Task Force (June 2010)</th><th>Center for a New American Security (CNAS) (November 2008)</th><th>Center for Strategic and Budgetary Assessments (CSBA) (2008)[b]</th></tr>
<tr><td colspan="9">Submarines</td></tr>
<tr><td>SSBN</td><td>12</td><td>7</td><td>14[c]</td><td>6</td><td>14</td><td>7</td><td>14</td><td>12</td></tr>
<tr><td>SSGN</td><td>0</td><td>6-7</td><td>4</td><td>0</td><td>4</td><td>4</td><td>0</td><td>2</td></tr>
<tr><td>SSN</td><td>48</td><td>42</td><td>55</td><td>40</td><td>55</td><td>37</td><td>40</td><td>41</td></tr>
<tr><td colspan="9">Aircraft carriers</td></tr>
<tr><td>CVN</td><td>11</td><td>9</td><td>11</td><td>8</td><td>11</td><td>9</td><td>8</td><td>11</td></tr>
<tr><td>CVE</td><td>0</td><td>0</td><td>0</td><td>0</td><td>0</td><td>0</td><td>0</td><td>4</td></tr>
<tr><td colspan="9">Surface combatants</td></tr>
<tr><td>Cruiser</td><td rowspan="2">88</td><td rowspan="2">72-74</td><td rowspan="2">88</td><td>22</td><td>n/a</td><td rowspan="2">85</td><td>18</td><td>14</td></tr>
<tr><td>Destroyer</td><td>65</td><td>n/a</td><td>56</td><td>73</td></tr>
<tr><td>Frigate</td><td>0</td><td>2-7[j]</td><td rowspan="2">28[d]</td><td>14</td><td>n/a</td><td>0</td><td>0</td><td>9[e]</td></tr>
<tr><td>LCS</td><td>52</td><td>12[j]</td><td>4</td><td>n/a</td><td>25</td><td>48</td><td>55</td></tr>
<tr><td>SSC</td><td>0</td><td>[j]</td><td>0</td><td>0</td><td>n/a</td><td>0</td><td>40</td><td>0[f]</td></tr>
<tr><td colspan="9">Amphibious and Maritime Prepositioning Force (Future) (MPF[F]) ships</td></tr>
<tr><td>Amphibious ships</td><td>33</td><td>≥23</td><td>37</td><td>23</td><td>n/a</td><td>27</td><td>36</td><td>33</td></tr>
<tr><td>MPF(F) ships</td><td>0</td><td>n/a</td><td>0</td><td>0</td><td>n/a</td><td>n/a</td><td>0</td><td>3[g]</td></tr>
<tr><td>LSD station ships</td><td>0</td><td>0</td><td>0</td><td>0</td><td>0</td><td>0</td><td>0</td><td>7[h]</td></tr>
<tr><td colspan="9">Other: Mine warfare (MIW) ships; Combat Logistics Force (CLF) ships (i.e., at-sea resupply ships), and support ships</td></tr>
<tr><td>MIW</td><td>0</td><td>14[j]</td><td>14</td><td>11</td><td>0</td><td>0</td><td>0</td><td>0</td></tr>
<tr><td>CLF ships</td><td>29</td><td>n/a</td><td>33</td><td>21</td><td>n/a</td><td>36</td><td>40</td><td>31</td></tr>
<tr><td>Support ships</td><td>33</td><td>n/a</td><td>25</td><td>27</td><td>n/a</td><td></td><td></td><td>31</td></tr>
<tr><td>TOTAL battle force ships</td><td>306</td><td>230</td><td>309</td><td>241</td><td>346</td><td>230</td><td>300</td><td>326[i]</td></tr>
</table>

Source: Table prepared by CRS based on the following sources: **For Heritage Foundation**: *A Strong National Defense[:] The Armed Forces America Needs and What They Will Cost*, Heritage Foundation,

April 5, 2011, pp. 25-26. **For Cato Institute**: Benjamin H. Friedman and Christopher Preble, *Budgetary Savings from Military Restraint*, Washington, Cato Institute, September 23, 2010 (Policy Analysis No. 667), pp. 6, 8-10, and additional information provided by Cato Institute to CRS by e-mail on September 22, 2010. **For Independent Panel Assessment**: Stephen J. Hadley and William J. Perry, co-chairmen, et al., *The QDR in Perspective: Meeting America's National Security Needs In the 21st Century, The Final Report of the Quadrennial Defense Review Independent Panel*, Washington, 2010, Figure 3-2 on pages 58-59. **For Sustainable Defense Task Force**: *Debt, Deficits, and Defense, A Way Forward[:] Report of the Sustainable Defense Task Force*, June 11, 2010, pp. 19-20. **For CNAS**: Frank Hoffman, *From Preponderance to Partnership: American Maritime Power in the 21st Century*. Washington, Center for a New American Security, November 2008. p. 19 (Table 2). **For CSBA**: Robert O. Work, *The US Navy[:] Charting a Course for Tomorrow's Fleet*. Washington, Center for Strategic and Budgetary Assessments, 2008. p. 81 (Figure 5). **For PDA**: Carl Conetta, Reasonable Defense, Project on Defense Alternatives, November 14, 2012, 31 pp.

Notes: **n/a** is not addressed in the report. **SSBN** is nuclear-powered ballistic missile submarine; **SSGN** is nuclear-powered cruise missile and special operations forces submarine; **SSN** is nuclear-powered attack submarine; **CVN** is large nuclear-powered aircraft carrier; **CVE** is medium-sized aircraft carrier; **LCS** is Littoral Combat Ship; **SSC** (an acronym created by CRS for this table) is small surface combatant of 1,000+ tons displacement—a ship similar to late-1990s Streetfighter concept; **MPF(F)** is Maritime Prepositioning Force (Future) ship; **LSD** is LSD-41/49 class amphibious ship operating as a station ship for a formation like a Global Fleet Station (GFS); **MIW** is mine warfare ship; **CLF** is combat logistics force (i.e., resupply) ship.

a. Figures shown are for the year 2020; for subsequent years, reductions from these figures would be considered.

b. Figures shown are for the year 2028.

c. The report calls for a force of 280 SLBMs, which appears to equate to a force of 14 SSBNs, each with 20 SLBM tubes.

d. The report calls for a force of 28 small surface combatants, and appears to use the term small surface combatants the same way that the Navy does in the 30-year shipbuilding plan—as a way of collectively referring to frigates and LCSs. The small surface combatants (SSCs) called for in the November 2008 CNAS report are separate from and smaller than the LCS.

e. Maritime Security Frigates.

f. Plan includes 28 patrol craft (PCs) of a few hundred tons displacement each, as well as 29 boat detachments and seven riverine squadrons.

g. Plan shows three Mobile Landing Platform (MLP) ships that the Navy currently plans for the MPF(F) squadron, plus 16 existing current-generation maritime prepositioning force (MPF) ships and 17 existing prepositioning ships for Army and other service/agency equipment. Plan also shows 67 other DOD sealift ships.

h. T-LSDs, meaning LSDs operated by the Military Sealift Command (MSC) with a partly civilian crew.

i. The CSBA report shows a total of 488 units by including 162 additional force units that do not count toward the 306-ship goal under the battle force ships counting method that has been used since the early 1980s for public policy discussions of the size of the Navy. These 162 additional force units include 16 existing current-generation maritime prepositioning force (MPF) ships and 17 existing prepositioning ships for Army and other service/agency equipment, 67 other DOD sealift ships, 28 PCs, 29 boat detachments, and certain other small-scale units. The CSBA report proposes a new counting method for naval/maritime forces that includes units such as these in the total count.

j. The report "prescribes ending procurement of the LCS with the 12 already purchased. The *Reasonable Defense* model foresees a future cohort of 28 to 33 small surface combatants, including a mix of the 12 LCS that have already been procured, 14 Mine Counter Measure (MCM) ships already in the fleet, and small frigates or ocean-going corvettes. As the MCM ships age and leave the fleet, the LCS should assume their role. The would leave a post-MCM requirement for 16 to 21 additional small surface combatants. For this, the Navy needs a simpler, less expensive alternative to the LCS."

The Navy's current goal for a fleet of 306 ships reflects a number of judgments and planning factors (some of which the Navy receives from the Office of the Secretary of Defense), including but not limited to the following:

- U.S. interests and the U.S. role in the world, and the U.S. military strategy for supporting those interests and that role;
- current and projected Navy missions in support of U.S. military strategy, including both wartime operations and day-to-day forward-deployed operations;
- current and projected capabilities of potential adversaries, including their anti-access/area-denial (A2/AD) capabilities;
- regional combatant commander (COCOM) requests for forward-deployed Navy forces;
- the individual and networked capabilities of current and future Navy ships and aircraft;
- basing arrangements for Navy ships, including numbers and locations of ships homeported in foreign countries;
- maintenance and deployment cycles for Navy ships; and
- fiscal constraints.

With regard to the fourth point above, Navy officials testified at least three times in 2012 that a Navy of more than 500 ships would be required to fully meet COCOM requests for forward- deployed Navy forces (see **Appendix A**). The difference between a fleet of more than 500 ships and the current goal for a fleet of 306 ships can be viewed as one measure of the operational risk associated with the goal of a fleet of 306 ships. A goal for a fleet of more than 500 ships might be viewed as a fiscally unconstrained goal.

Some study groups have made their own proposals for Navy ship force structure that reflect their own perspectives on the points listed above (particularly the first three and the final one) shows some of these proposals. For purposes of comparison, also shows the Navy's 306-ship goal of January 2013.

A potential key question for Congress concerns whether the U.S. Navy in coming years will be large enough to adequately counter improved Chinese maritime A2/AD forces while also adequately performing other missions of interest to U.S. policymakers around the world. Some observers are concerned that a combination of growing Chinese naval capabilities and budget- driven reductions in the size of the U.S. Navy could encourage Chinese military overconfidence and demoralize U.S. allies and partners in the Pacific, and thereby make it harder for the United States to defend its interests in the region.[41] Potential oversight questions for Congress include the following:

- Under the Administration's plans, will the Navy in coming years be large enough to adequately counter improved Chinese maritime A2/AD forces while also adequately performing other missions of interest to U.S. policymakers around the world?
- What might be the political and security implications in the Asia-Pacific region of a combination of growing Chinese naval capabilities and budget-driven reductions in the size of the U.S. Navy?

- If the Navy is reduced in size and priority is given to maintaining Navy forces in the Pacific, what will be the impact on Navy force levels in other parts of the world, such as the Persian Gulf/Indian Ocean region or the Mediterranean Sea, and consequently on the Navy's ability to adequately perform its missions in those parts of the world?
- To what extent could the operational impacts of a reduction in Navy ship numbers be mitigated through increased use of forward homeporting, multiple crewing, and long-duration deployments with crew rotation (i.e., "Sea Swap")? How feasible are these options, and what would be their potential costs and benefits?[42]
- Particularly in a situation of constrained DOD resources, if enough funding is allocated to the Navy to permit the Navy in coming years to maintain a fleet of 306 ships including 11 aircraft carriers, how much would other DOD programs need to be reduced, and what would be the operational implications of those program reductions in terms of DOD's overall ability to counter improved Chinese military forces and perform other missions?[43]

Sufficiency of 30-Year Shipbuilding Plan for Achieving Navy's 306-Ship Goal

Another potential oversight issue for Congress concerns the sufficiency of the Navy's 30-year shipbuilding plan for achieving the Navy's goal for a 306-ship fleet. The Navy's 30-year shipbuilding plans in recent years have generally not included enough ships to fully support all elements of the Navy's 306-ship goal over the long run. The Navy has projected that the fleet would remain below 306 ships during most of the 30-year period, and experience shortfalls at various points in cruisers-destroyers, attack submarines, and amphibious ships. In light of these projected shortfalls, policymakers may wish to consider various options, including but not limited to the following:

- increasing planned procurement quantities of destroyers, attack submarines, and amphibious ships;
- extending, if feasible, the service lives of older destroyers and amphibious ships, and (if feasible) refueling and extending the service lives of a small number of older attack submarines; and
- reducing the Navy's ship force structure goals for destroyers, attack submarines, and/or amphibious ships.

The first two options would require increased funding for procurement and for operation and support (O&S) costs, respectively, while the third option could reduce the Navy's ability to perform its missions.

Affordability of 30-Year Shipbuilding Plan

Another potential oversight issue for Congress concerns the prospective affordability of the Navy's 30-year shipbuilding plan. In assessing the prospective affordability of the 30-year plan, key factors that Congress may consider include estimated ship procurement costs and future shipbuilding funding levels.

Estimated Ship Procurement Costs

As mentioned earlier, the Navy's 30-year shipbuilding plan is based on certain assumptions, including assumptions about ship procurement costs. If one or more Navy ship designs turn out to be more expensive to build than the Navy estimates, then the projected funding levels shown in the 30-year shipbuilding plan will not be sufficient to procure all the ships shown in the plan. Ship designs that can be viewed as posing a risk of being more expensive to build than the Navy estimates include Gerald R. Ford (CVN-78) class aircraft carriers (a program currently experiencing cost growth), Ohio-replacement (SSBNX) class ballistic missile submarines, the Flight III version of the DDG-51 destroyer, the TAO(X) oiler, and the LX(R) amphibious ship.

In recent years, the Congressional Budget Office (CBO) has estimated that certain Navy ships would be more expensive to procure than the Navy estimates, and consequently that the Navy's 30-year shipbuilding plan would cost more to implement than the Navy has estimated. In its October 2013 report on the cost of the FY2014 30-year shipbuilding plan, the CBO estimates that the plan would cost an average of $19.3 billion per year in constant FY2013 dollars to implement, or about 15% more than the Navy estimates. CBO's estimate is about 6% higher than the Navy's estimate for the first 10 years of the plan, about 14% higher than the Navy's estimate for the second 10 years of the plan, and about 26% higher than the Navy's estimate for the final 10 years of the plan.[44] Some of the difference between CBO's estimate and the Navy's estimate, particularly in the latter years of the plan, is due to a difference between CBO and the Navy in how to treat inflation in Navy shipbuilding. **Table 8** summarizes the Navy and CBO estimates of the FY2014 30-year shipbuilding plans.

Future Shipbuilding Funding Levels

As mentioned earlier, the Navy's 30-year shipbuilding plan is based on certain assumptions, including assumptions about future shipbuilding funding levels. It has been known for some time that funding requirements for the Ohio-replacement (SSBN[X]) ballistic missile submarine program will put considerable pressure on the shipbuilding budget during the middle years of the 30-year plan. Although the FY2014 30-year shipbuilding plan reduces procurement of other types of ships in the middle years of the plan to help accommodate the SSBN(X) program, the Navy still projects that the shipbuilding budget would need to be substantially higher during the middle years of the plan than during the earlier or later years of the plan.

If the "hump" in shipbuilding funding during the middle years of the 30-year plan is not achieved, numerous ships shown for procurement during the middle years of the plan might not be procured. A potential oversight question for Congress is whether the Navy has received a commitment or assurance of some kind from DOD leaders that the Navy will be able to budget for the "hump" in shipbuilding funding during the middle years of the 30-year plan without reducing funding for other Navy program priorities. The Navy's report on the FY2014 30-year shipbuilding plan states:

Table 8. Navy and CBO Estimates of Cost of FY2014 30-Year (FY2014-FY2043) Shipbuilding Plans
Funding for new-construction ships, in billions of constant FY2013 dollars

	First 10 years (FY2014-FY2023)	Next 10 years (FY2024-FY2033)	Final 10 years (FY2034-FY2043)	Entire 30 years (FY2014-FY2043)
Navy estimate	15.4	19.8	15.2	16.8
CBO estimate	16.3	22.6	19.1	19.3
% difference between Navy and CBO estimates	6%	14%	26%	15%

Source: Congressional Budget Office, *An Analysis of the Navy's Fiscal Year 2014 Shipbuilding Plan*, October 2013, Table 3 (page 13).

The Department [of the Navy] will encounter several challenges in executing this shipbuilding plan; perhaps the most important is funding and delivering the *Ohio*-replacement (OR) program SSBN. The OR SSBN is projected to cost about $6 billion each [in constant FY2013 dollars]. Therefore, during the procurement and construction of OR SSBN between FY2021 and FY2035 an average of $19.2 billion per year is projected to be required for shipbuilding, which will be a key resourcing challenge for the Department.

In addition to the challenge of funding the OR SSBN, during several years in the early 20202 [the] Navy will also require approximately $2 billion [per year] in additional ship construction funding to recapitalize the large number of ships decommissioning in those years. Our current fleet has a large cohort of ships that are about the same age and will thus retire as a group. These ships were built in the 1980s, some at a rate of three or four ships per year per class. These retiring ships will need to be recapitalized to reach the FSA-required battle force size and shape [i.e., the 306-ship goal]....

DoN has historically been able to resource between $12B [billion] and $14B in annual new- ship procurement funding. During the FY2014-FY2018 FYDP, average annual new-ship procurement funding is about $14B. This level of investment is based on the need to balance our resources between manning, maintenance, sustainment, modernization and recapitalization of our ships, aircraft and weapons.

The cost of the OR SSBN is significant relative to the resources available to DoN in any given year. At the same time, the Department will have to address the block retirement of ships procured in large numbers during the 1980s which are reaching the end of their service lives. The confluence of these events prevents DoN from being able to shift resources within the shipbuilding account to accommodate the cost of the OR SSBN.

If DoN funds the OR SSBN from within its own resources, OR SSBN construction will take away from construction of other ships in the battle force such as attack submarines, destroyers, aircraft carriers and amphibious warfare shps. The resulting battle force will not meet the requirements of the FSA and will therefore not be sufficient to implement the DSG [Defense Strategic Guidance]. In addition there will be significant impact to the shipbuilding industrial base.[45]

A May 2, 2013, press report states:

[Vice Admiral William Burke, Deputy Chief of Naval Operations for Warfare Systems], who is set to retire in the next few weeks,[46] spoke frankly about the undersea

> portion of the U.S. strategic nuclear triad "and its intersection with our shipbuilding plan."
>
> His conclusion: "If we buy the [the 12 planned Ohio replacement (SSBNX) ballistic missile submarines] within existing [Navy] funds, we will not reach 300 ships. In fact, we'll find ourselves closer to 250. At these numbers, our global presence will be reduced such that we'll only be able to visit some areas of the world episodically."
>
> Sequestration will only make the situation worse. Burke said it would cause the Navy "to both reduce procurement as well as retire existing ships, leaving us with a Navy in the vicinity of 200 ships, at which point we may not be considered a global navy."[47]

In a situation of reduced levels of defense spending, such as what would occur if defense spending were to remain constrained to the revised cap levels in the Budget Control Act, the affordability challenge posed by the 30-year shipbuilding plan would be intensified. Even then, however, the current 30-year shipbuilding plan would not necessarily become unaffordable.[48]

The Navy estimates that, in constant FY2013 dollars, fully implementing the current 30-year shipbuilding plan would require an average of $16.8 billion in annual funding for new-construction ships, compared to a historic average of $12 billion to $14 billion provided for this purpose.[49] The required increase in average annual funding of $2.8 billion to $4.8 billion per year equates to less than 1% of DOD's annual budget under the revised caps of the Budget Control Act. CBO estimates that, in constant FY2013 dollars, fully implementing the current 30-year shipbuilding plan would require an average of $19.3 billion in annual funding for new- construction ships, or $2.5 billion per year more than the Navy estimates.[50] This would make the required increase in average annual funding $5.3 billion to $7.3 billion per year, which equates to roughly 1.1% to 1.5% of DOD's annual budget under the revised caps of the Budget Control Act.

Some observers, noting the U.S. strategic rebalancing toward the Asia-Pacific region, have advocated shifting a greater share of the DOD budget to the Navy and Air Force, on the grounds that the Asia-Pacific region is primarily a maritime and aerospace theater for DOD. In discussing the idea of shifting a greater share of the DOD budget to the Navy and Air Force, some of these observers refer to breaking the so-called "one-third, one-third, one-third" division of resources among the three military departments—a shorthand term sometimes used to refer to the more-or- less stable division of resources between the three military departments that existed for the three decades between the end of U.S. participation in the Vietnam War in 1973 and the start of the Iraq War in 2003.[51] In a context of breaking the "one-third, one-third, one-third" allocation with an aim of better aligning defense spending with the strategic rebalancing, shifting 1.5% or less of DOD's budget into the Navy's shipbuilding account would appear to be quite feasible.

More broadly, if defense spending were to remain constrained to the revised cap levels in the Budget Control Act, then fully funding the Department of the Navy's total budget at the levels shown in the current Future Years Defense Plan (FYDP) would require increasing the Department of the Navy's share of the non-Defense-Wide part of the DOD budget to about 41%, compared to about 36% in the FY2014 budget and an average of about 37% for the three-decade period between the Vietnam and Iraq wars.[52] While shifting 4% or 5% of DOD's budget to the Department of the Navy would be a more ambitious reallocation than shifting 1.5% or less of the DOD budget to the Navy's shipbuilding account, similarly large reallocations have occurred in the past:

- From the mid-1950s to the mid-1960s, reflecting a U.S. defense strategy at the time that placed a strong reliance on the deterrent value of nuclear weapons, the Department of the Air Force's share of the non-Defense-Wide DOD budget increased by several percentage points. The Department of the Air Force's share averaged about 45% for the 10-year period FY1956-FY1965, and peaked at more than 47% in FY1957-FY1959.
- For the 11-year period FY2003-FY2013, as a consequence of combat operations in Iraq and Afghanistan, the Department of the Army's share of the non-Defense- Wide DOD budget increased by roughly 10 percentage points. The Department of the Army's share during this period averaged about 39%, and peaked at more than 43% in FY2008. U.S. combat operations in Iraq and Afghanistan during this period reflected the implementation of U.S. national strategy as interpreted by policymakers during those years.

The point of the foregoing is not to argue whether it would be right or wrong to shift more of the DOD budget to the Navy's shipbuilding account or to the Department of the Navy's budget generally. Doing that would require reducing funding for other DOD programs, and policymakers would need to weigh the resulting net impact on overall DOD capabilities. The point, rather, is to note that the allocation of DOD resources is not fixed, that aligning DOD spending with U.S. strategy in coming years could involve changing the allocation by more than a very marginal amount, and that such a changed allocation could provide the funding needed to implement the current 30-year shipbuilding plan. The alternative of assuming at the outset that there is no potential for making anything more than very marginal shifts in the allocation of DOD resources could unnecessarily constrain options available to policymakers and prevent the allocation of DOD resources from being aligned optimally with U.S. strategy.

As an alternative or supplement to the option of altering the allocation of DOD resources among the military departments, the 30-year shipbuilding plan could also become more affordable by taking actions beyond those now being implemented by DOD to control military personnel pay and benefits and reduce what some observers refer to as DOD's overhead or back-office costs. Multiple organizations have made recommendations for such actions in recent years. The Defense Business Board, for example, estimated that at least $200 billion of DOD's enacted budget for FY2010 constituted overhead costs. The board stated, "There has been an explosion of overhead work because the Department has failed to establish adequate controls to keep it in line relative to the size of the warfight," and that "in order to accomplish that work, the Department has applied ever more personnel to those tasks which has added immensely to costs." The board stated further that "whether it's improving the tooth-to-tail ratio; increasing the 'bang for the buck', or converting overhead to combat, Congress and DoD must significantly change their approach," and that DOD "must use the numerous world-class business practices and proven business operations that are applicable to DoD's overhead."[53]

One potential way to interpret the affordability challenge posed by the Navy's 30-year shipbuilding plan is to view it as an invitation by the Navy for policymakers to consider matters such as the alignment between U.S. strategy and the division of DOD resources among the military departments, and the potential for taking actions beyond those now being implemented by DOD to control military personnel pay and benefits and reduce DOD overhead and back- office costs. The Navy's prepared statement for a September 18, 2013, hearing before the House Armed Services Committee on planning for sequestration in FY2014 and the perspectives of the military services on the Strategic Choices and Management Review (SCMR) provides a number of details about reductions in Navy force structure and acquisition programs that could result from constraining DOD's budget to the revised cap levels in the Budget Control Act.[54] These potential reductions do not appear to reflect any substantial shift in the allocation of DOD resources among the military departments, or the taking of actions beyond those already being implemented by DOD to control DOD personnel pay and benefits and reduce DOD overhead and back-office costs. The fact that the Navy in its prepared statement did not choose to discuss the possibility of a changed allocation of DOD resources among the military departments or additional actions to control DOD personnel pay and benefits and reduce DOD overhead and back-office costs does not prevent Congress from considering such possibilities.

LEGISLATIVE ACTIVITY FOR FY2014

FY2014 Funding Request

The Navy's proposed FY2014 budget requests funding for the procurement of 8 new battle force ships (i.e., ships that count against the 306-ship goal)—two Virginia-class attack submarines, one DDG-51 class Aegis destroyer, four Littoral Combat Ships (LCSs), and one Mobile Landing Platform/Afloat Forward Staging Base (MLP/AFSB) ship. These ships are all funded through the Shipbuilding and Conversion, Navy (SCN) account.

FY2014 National Defense Authorization Act (H.R. 1960/S. 1197)

House

The House Armed Services Committee, in its report (H.Rept. 113-102 of June 7, 2013) on H.R. 1960, supported the Navy's requests for funding for the procurement of 8 new battle force ships, approving, and in some cases recommending increases above, the requested amounts. (Pages 387-388) The report recommended:

- an increase of $492 million for the Virginia-class attack submarine program, which almost matches the $492.3 million sequestered from the program by the March 1, 2013, sequester (see "Impact of March 1, 2013, Sequester on FY2013 Funding" in "Oversight Issues for Congress for FY2014");
- an increase of $332 million for the DDG-51 program, which is similar to the amount of additional funding that the Navy has testified would be needed to execute the third DDG-51 procured in FY2013 (i.e., the tenth DDG-51 in the FY2013-FY2017 DDG-51 multi-year procurement contract);
- an increase of $79.3 million for the DDG-1000 destroyer program, which is $9 million more than the $70.3 million sequestered from the program by the March 1, 2013, sequester;

- an increase of $23.4 million for the Moored Training Ship program, matching the amount sequestered from the program by the March 1, 2013, sequester; and
- an increase of $7.6 million for the JHSV program, which is about one-third as much as the $21.4 million sequestered from the program by the March 1, 2013, sequester.

Section 1022 of H.R. 1960 as reported states:

SEC. 1022. AVAILABILITY OF FUNDS FOR RETIREMENT OR INACTIVATION OF TICONDEROGA CLASS CRUISERS OR DOCK LANDING SHIPS.

(a) Limitation on Availability of Funds-

(1) IN GENERAL- Except as provided in paragraph (2), none of the funds authorized to be appropriated by this Act or otherwise made available for fiscal year 2014 for the Department of Defense may be obligated or expended to retire, prepare to retire, inactivate, or place in storage a cruiser or dock landing ship.

(2) EXCEPTION- Notwithstanding paragraph (1), the funds referred to in such subsection may be obligated or expended to retire the U.S.S. Denver, LPD9.

(b) Authority to Transfer Authorizations-

(1) AUTHORITY- Subject to the availability of appropriations for such purpose, the Secretary of Defense may transfer amounts of authorizations made available to the Department of Defense for fiscal year 2013 specifically for the modernization of vessels referred to in subsection (a)(1). Amounts of authorizations so transferred shall be merged with and be available for the same purposes as the authorization to which transferred.

(2) LIMITATION- The total amount of authorizations that the Secretary may transfer under the authority of this subsection may not exceed $914,676,000.

(3) ADDITIONAL AUTHORITY- The transfer authority provided by this subsection is in addition to the transfer authority provided under section 1001 of this Act and under section 1001 of the National Defense Authorization Act for Fiscal Year 2013 (P.L. 112-239; 126 Stat. 1902).

Regarding Section 1022, H.Rept. 113-102 states:

Section 1022—Availability of Funds for Retirement or Inactivation of Ticonderoga Class Cruisers or Dock Landing Ships

This section would limit the obligation and expenditure of funds authorized to be appropriated or otherwise made available for fiscal year 2014 for the retirement, inactivation, or storage of a cruiser or dock landing ship. This section would provide an exception for the retirement of the U.S.S. Denver (LPD 9).

This section would further provide for transfer authority for the purpose of providing sufficient appropriations to support the modernization of seven cruisers. If requested by the Secretary of Defense, the committee believes the following transfers should be included: OPN Line 0960, $662.7 million; OPN Line 2312, $1.8 million; OPN Line 2360, $6.6 million; OPN Line 2915, $13.7 million; OPN Line 3050, $13.4 million; OPN Line 3216, $20.8 million; OPN Line 5530, $4.6 million; WPN Line 4223, $91.1 million; and RDTE Line 1447, $100.0 million. The total transfer authority is $914.7 million. (Pages 233-234)

Section 1024 states:

SEC. 1024. SENSE OF CONGRESS REGARDING A BALANCED FUTURE NAVAL FORCE.

(a) Findings- Congress makes the following findings:
   (1) The battle force of the Navy must be sufficiently sized and balanced in capability to meet current and anticipated future national security objectives.
   (2) A robust and balanced naval force is required for the Department of Defense to fully execute the President's National Security Strategy.
   (3) To develop and sustain required capabilities the Navy must balance investment and maintenance costs across various ship types, including—
      (A) aircraft carriers;
      (B) surface combatants;
      (C) submarines;
      (D) amphibious assault ships; and
      (E) other auxiliary vessels, including support vessels operated by the Military Sealift Command.
   (4) Despite a Marine Corps requirement for 38 amphibious assault ships, the Navy possesses only 30 amphibious assault ships with an average of 22 ships available for surge deployment.
   (5) The inadequate level of investment in Navy shipbuilding over the last 20 years has resulted in—
      (A) a fragile shipbuilding industrial base, both in the construction yards and secondary suppliers of materiel and equipment; and
      (B) increased costs per vessel stemming from low production volume.
   (6) The Department of Defense, Military Construction and Veterans Affairs, and Full-Year Continuing Appropriations Act for Fiscal Year 2013 provided $263,000,000 towards the advance procurement of materiel and equipment required to continue the San Antonio LPD 17 amphibious transport dock class to a total of 12 ships, a key first step in rebalancing the amphibious assault ship force structure.
(b) Sense of Congress- It is the Sense of Congress that—
   (1) the Department of Defense and the Department of the Navy must prioritize funding towards increased shipbuilding rates to enable the Navy to meet the full-range of combatant commander requests;
   (2) the Department of the Navy's future budget requests and the Long Range Plan for the Construction of Naval Forces must realistically anticipate and reflect the true investment necessary to meet stated force structure goals;
   (3) without modification to Long Range Plan for the Construction of Naval Forces shipbuilding plan, the future of the industrial base that enables construction of large, combat- survivable amphibious assault ships is at significant risk; and
   (4) the Department of Defense and Congress should act expeditiously to restore the force structure and capability balance of the Navy fleet as quickly as possible.

Regarding Section 1024, H.Rept. 113-102 states:

Section 1024—Sense of Congress Regarding a Balanced Future Naval Force

This section would provide the Sense of Congress that additional funding should be prioritized toward shipbuilding efforts and that Department of Navy budget projections should realistically anticipate the true investment to meet force structure goals. (Page 234)

Section 1233 states:

SEC. 1233. SENSE OF CONGRESS ON THE DEFENSE OF THE ARABIAN GULF.

(a) Findings- Congress finds the following:
   (1) In response to U.S. Central Command requirements, the United States Navy has maintained, on average, more than one aircraft carrier in the Arabian Gulf for more than five years.
   (2) In February 2013, the senior leadership of the Department of Defense elected to reduce the number of aircraft carriers deployed to the Arabian Gulf in light of budget constraints and limitation of the overall carrier force structure to support the two aircraft carrier requirement.
   (3) In reference to the decision to indefinitely delay the deployment of the USS Harry Truman, CVN 75, and the USS Gettysburg, its cruiser escort, Chairman of the Joint Chiefs, General Martin Dempsey stated, `We're trying to stretch our readiness out by keeping this particular carrier in homeport in our global response force, so if something happens elsewhere in the world, we can respond to it. Had we deployed it and `consumed' that readiness, we could have created a situation where downstream we wouldn't have a carrier present in certain parts of the world at all.'.
   (4) Highlighting the risks of having only one aircraft carrier in the region and relying on land-based aircraft, General Dempsey stated, `When you have carrier-based aircraft, you have complete autonomy and control over when you use them. When you use land-based aircraft, you often have to have host-nation permission to use them.'.
   (5) Addressing the perception of the United States commitment to the region, General James Mattis, Commander of U.S. Central Command, testified in March 2013, `Perhaps the greatest risk to U.S. interests in the region is a perceived lack of an enduring U.S. commitment to collective interests and the security of our regional partners.'. He went on to testify that, `The drawdown of our forces can be misinterpreted as a lack of attention, a lack of commitment to the region.'.

(b) Sense of Congress- It is the sense of Congress that—
   (1) maintaining only one aircraft carrier battle group in the Arabian Gulf constrains United States' options and could put at risk the ability to have diversified platforms from which to defend the Arabian Gulf and, if necessary, to conduct military operations to prevent Iran from threatening the United States, United States allies, or Iran's neighbors with nuclear weapons;
   (2) it is in the interests of the United States to maintain both land-based and sea-based capabilities in the region to project force;

(3) land-based locations in the region could restrict United States military options and critically impact the operational capability if required to conduct a defense of the Arabian Gulf because the United States has not finalized bilateral security agreements with key Gulf Cooperation Council countries;
(4) as a result of these and other critical limitations associated with maintaining one aircraft carrier battle group in the Arabian Gulf, United States military commanders have expressed concerns about the operational constraints, the increasing uncertainty among United States allies, and the emboldening of potential adversaries such as Iran;
(5) regarding the ability of the United States Navy to maintain a two aircraft carrier presence in the Arabian Gulf, the Chief of Naval Operations, Admiral Jonathan Greenert, stated, `We need 11 carriers to do the job. That's been pretty clearly written, and that's underwritten in our defense strategic guidance.'.
(6) the United States should construct and sufficiently sustain a fleet of at least eleven aircraft carriers and associated battle force ships in order to meet current and future requirements and to support at least a two aircraft carrier battle group presence in the Arabian Gulf, in addition to meeting other operational requirements; and
(7) the United States should finalize bilateral agreements with key Gulf Cooperation Council countries that support the Defense of the Arabian Gulf requirements, at the earliest possible date.

Regarding Section 1233, H.Rept. 113-102 states:

Section 1233—Sense of Congress on the Defense of the Arabian Gulf

This section would express the sense of Congress with respect to the United States' operational posture and capacity to defend the Arabian Gulf, including the risk of maintaining only 1 aircraft carrier battle group in the Arabian Gulf to deter the Government of the Islamic Republic of Iran, and the limitations on maintaining a 2 aircraft carrier battle group presence in the Arabian Gulf, stemming in part from not constructing and sustaining a fleet of at least 11 aircraft carriers. Additionally, this section would express the sense of Congress that the United States should finalize bilateral security agreements with key Gulf Cooperation Council countries that support the Defense of the Arabian Gulf requirements at the earliest possible date. (Page 273)

H.Rept. 113-102 also states:

The committee is concerned about the Navy's overall fleet size and the continuous sustained demand for naval forces, especially in light of the Administration's strategic shift to operations in the Asia-Pacific. Therefore, the restriction precluding the Navy from retiring seven Ticonderoga-class guided missile cruisers and two amphibious ships well before the end of their expected service life continues for fiscal year 2014. The committee would provide additional funds to the Navy to properly modernize and maintain these critical naval assets. The committee notes that it is less costly to maintain existing assets than to procure new ones and this funding ensures the correct naval capabilities and fleet mix for the length of time originally authorized by Congress....

The committee also would fund needed ship construction to obviate the negative consequences of sequestration on the various ship construction programs. The committee would provide sufficient funds to support the acquisition of the 10th DDG–51 class destroyer of a multiyear procurement; additional funds to support the continued acquisition of two Virginia Class attack submarines; and additional funds associated with the completion of the DDG–1000 class destroyer, the Moored Training Ship, and the Joint High Speed Vessel. (Page 6)

H.Rept. 113-102 also states:

Long-range plan for the construction of naval vessels

Pursuant to section 231 of title 10, United States Code, the Secretary of Defense provided the annual long-range plan for the construction of naval vessels on May 10, 2013, as informed by the Future Years Defense Program (FYDP) for fiscal years 2014–18. The Secretary also indicated that a force structure of "about 300 ships" would be necessary to support ongoing naval operations. The Secretary further highlights the "resourcing challenges outside the FYDP largely due to investment requirements associated with the SSBN(X) program". The Secretary acknowledges that these ship construction pressures will precipitate higher fiscal requirements in the mid-term planning period (fiscal years 2024–33) requiring an annual investment of $19.8 billion per year in fiscal year 2013 constant dollars.

The committee supports a robust shipbuilding plan that invests in the near and long term needs of our Navy, and considers the recapitalization of the SSBN fleet a challenging but necessary strategic priority. However, the committee is concerned that the Navy's ship construction accounts will face significant pressure in supporting long term ship requirements while also resourcing the Ohio-class replacement ballistic missile submarine program. The committee also believes that a significant increase to the ship construction accounts is unsustainable in times of budget challenges.

The Congressional Budget Office has estimated that the average ship construction investment over the last 30 years, in current dollars, is $16.0 billion. Therefore, to better understand the significance associated with even sustaining the current ship construction investment throughout the long-range plan, the committee directs the Secretary of the Navy to provide a report to the congressional defense committee by March 1, 2014, that provides an update to the long plan for the construction of naval vessels based on $16.0 billion across the entirety of the long-range plan and to assess the corresponding reductions in the shipbuilding plan. The Secretary of the Navy should also provide an assessment of this investment in terms of the health associated with the industrial base, as well as a discussion of alternative strategies for the Navy and Congress to consider in alleviating any shortfalls between this assessment and the May 10 report. (Page 29)

Senate

The Senate Armed Services Committee, in its report (S.Rept. 113-44 of June 20, 2013) on S. 1197, supported the Navy's requests for funding for the procurement of 8 new battle force ships, approving, and in two cases recommending increases above, the requested amounts. (Pages 295-296) The report recommended:

- an increase of $100 million for the DDG-51 program, to "Help buy [a] 3rd DDG– 51 in FY[20]13;" and
- an increase of $55.3 million for the Afloat Forward Staging Base (AFSB) program, reflecting a "Navy requested adjustment."

Section 1021 of S. 1197 as reported states:

SEC. 1021. MODIFICATION OF REQUIREMENTS FOR ANNUAL LONG-RANGE PLAN FOR THE CONSTRUCTION OF NAVAL VESSELS.

(a) Annual Naval Vessel Construction Plan- Subsection (b) of section 231 of title 10, United States Code, is amended—
  (1) in paragraph (1)—
    (A) by striking `should be designed' both places it appears and inserting `shall be designed'; and
    (B) by striking `is capable of supporting' both places it appears and inserting `supports'; and
  (2) in paragraph (2)—
    (A) in subparagraph (B), by inserting `and capabilities' after `naval vessel force structure'; and
    (B) by adding at the end the following new subparagraph:
      `(D) The estimated total cost of construction for each vessel used to determine estimated levels of annual funding under subparagraph (C).'.

(b) Assessment When Construction Plan Does Not Meet Force Structure Requirements- Such section is further amended—
  (1) by redesignating subsections (d), (e), and (f) as subsections (e), (f), and (g), respectively; and
  (2) by inserting after subsection (c) the following new subsection (d):
    `(d) Assessment When Annual Naval Vessel Construction Plan Does Not Meet Force Structure Requirements- If the annual naval vessel construction plan for a fiscal year under subsection (b) does not result in a force structure or capabilities that meet the requirements identified in subsection (b)(2)(B), the Secretary shall include with the defense budget materials for that fiscal year an assessment of the extent of the strategic and operational risk to national security associated with the reduced force structure of naval vessels over the period of time that the required force structure or capabilities are not achieved. Such assessment shall include an analysis whether the risks are acceptable, and plans to mitigate such risks. Such assessment shall be coordinated in advance with the commanders of the combatant commands and the Nuclear Weapons Council under section 179 of this title.'.

Regarding Section 1021, S.Rept. 113-44 states:

Modification of requirements for annual long-range plan for the construction of naval vessels (sec. 1021)

The committee recommends a provision that would modify section 231 of title 10, Unites Sates Code, to include a requirement to report on the total cost of construction for each vessel used to determine estimated levels of annual funding in the report, and an assessment of the extent of the strategic and operational risk to national security whenever the number or capabilities of the naval vessels in the plan do not meet requirements. (Page 172)

Section 1022 states:

SEC. 1022. REPORT ON NAVAL VESSELS AND THE FORCE STRUCTURE ASSESSMENT.

(a) Report Required- Not later than February 1, 2014, the Chief of Naval Operations shall submit to the congressional defense committees a report on current and anticipated requirements for combatant vessels of the Navy over the next 30 years.
(b) Elements- The report required by subsection (a) shall include the following:
  (1) A description of the naval capability requirements identified by the combatant commands in developing the Force Structure Assessment (FSA) in 2005 and revalidating that Assessment in 2010.
  (2) The capabilities for each class of vessel that was assumed in the Force Structure Assessment.
  (3) An assessment of the capabilities of the current fleet of combatant vessels of the Navy to meet current and anticipated requirements.
  (4) An assessment the capabilities of the anticipated fleet of combatant vessels of the Navy to meet emerging threats over the next 30 years.
  (5) An assessment of how the Navy will meet combatant command requirements for forward-deployed naval capabilities with a smaller number of ships and submarines.
  (6) An assessment of how the Navy will manage the risk of massing a greater set of capabilities on a smaller number of ships while facing an expanding range of asymmetrical threats, such as—
    (A) anti-access/area-denial capabilities;
    (B) diesel-electric submarines;
    (C) mines; and
    (D) anti-ship cruise and ballistic missiles.
(c) Form- The report required by subsection (a) shall be submitted in unclassified form, but may include a classified annex.

Regarding Section 1022, S.Rept. 113-44 states:

Report on naval vessels and the Force Structure Assessment (sec. 1022)

The committee recommends a provision that would direct the Chief of Naval Operations (CNO) to provide a report to the congressional defense committees no later than February 1, 2014, that would assess the current fleet capabilities compared to the threat and the likely situation over the next 30 years. The CNO should produce an unclassified report, and a classified annex to that report.

Section 1105 of the National Defense Authorization Act for Fiscal Year 2013 (Public Law 112–239) required the Secretary of the Navy to provide a comprehensive description of the current requirements for combatant vessels of the Navy, including submarines. The Navy submitted that report which reflected that, while the previous requirement was for a fleet of 313 ships, originally identified in a 2005 Force Structure Assessment (FSA) and revalidated in 2010, the new requirements have declined to a total of 306 ships.

Within this seven-ship net decrease to 306 ships, actual combat power is reduced by six large surface combatants, three small surface combatants, and four cruise missile submarines, for a total reduction of 13 combatant ships, offset by an increase of six command and support ships. The reduction in cruise missile submarines reflects a Navy plan not to replace these ships when they retire in the late 2020s.

In congressional testimony this year about the 2012 FSA, representatives of the Department of the Navy asserted that the revised fleet size goal considered various factors, including the revised Defense Strategic Guidance in 2012, changed Navy mission requirements, different individual ship capabilities, fleet networking capabilities, and ship basing arrangements and operational cycles, as opposed to just platforms or numbers of ships.

The committee notes that the current fleet of 283 battle force ships provides combatant commanders (COCOM) with a greater array of fire power than did the Navy fleets of 500 or more ships of the Cold War era, with precision-guided air-delivered weapons, numbers of Tomahawk-capable ships, and the sophistication of command, control, communications, and computer systems; intelligence, surveillance, and reconnaissance (ISR) systems; and networking capabilities that did not exist during the Cold War.

The President's 2012 revised U.S. defense strategy calls for a rebalancing towards the Asia- Pacific region, a predominantly maritime and aerospace domain. This will most likely increase demands for Navy fleet resources and deployments in that region in competition with global demand. Regarding global demands for naval forces, Navy officials testified last year that fully satisfying COCOM requests for forward-deployed Navy forces in various regions would require more than 500 ships.

The Report to Congress on the Annual Long-Range Plan for Construction of Naval Vessels for FY 2014, dated May 10, 2013, proposes near-term retirements of cruisers and amphibious ships, resulting in 270 ships, the smallest fleet since 1917. The committee is concerned that the plan will not meet the goal of a 306-ship battle force inventory until 2037 and assumes risk over the 30-year period, with periodic shortages of aircraft carriers, cruisers, destroyers, attack submarines, and amphibious ships. The committee notes that the Navy possesses only 28 amphibious ships, with an average of only 22 ships available for surge deployment, despite a Marine Corps requirement for 38 amphibious ships. As such, the committee is also concerned that the Navy's shortfall in amphibious ships adds risk to the Marine Corps' ability to meet current and future COCOM requests.

There are also risks with the Navy's plan beyond mere numbers of ships:

(1) within the plan, the Navy intends that the Littoral Combat Ship (LCS) will comprise over one third of the Navy's total surface combatant fleet by 2028. This fact compounds risk, since the LCS to date has not completed operational testing or demonstrated adequate performance of assigned missions in critical areas of mine countermeasures, anti-surface warfare, or anti-submarine operations;

(2) with its decision in the fiscal year 2013 budget to delay the procurement of the first Ohio- class replacement ballistic missile submarine by 2 years, to fiscal year 2021, the Navy has consumed all schedule margin for replacing the existing Ohio-class boats on a timely basis, and increased the risk that an unforeseen event affecting a strategic missile submarine's operational availability would prevent the Navy from being able to fully satisfy U.S. Strategic Command (STRATCOM) requirements; and

(3) the 2012 FSA and the shipbuilding plan do not account for the possible effects of further budget cuts under the Budget Control Act of 2011. Various representatives of the Navy have testified that, if Department of Defense budgets are reduced below levels shown in the President's budget request for fiscal year 2014, the Navy will have to reduce the size and capabilities of the fleet to reflect further changes in U.S. defense strategy and available Navy resources.

Given the risks inherent in the planned size of the Navy, the capabilities of the ships planned, the rate of Navy ship procurement, and the potential affordability of the

Navy's shipbuilding plans at a time of significant uncertainty in defense budgets, the committee agrees that in order to adequately assess the ability of the fleet proposed by the 2012 FSA to meet national security requirements, there needs to be an objective set of standards to assess the ability of the Navy to meet national security objectives, as opposed to simply relying on numbers and types of platforms.

The committee also believes that a failure to recapitalize our seabased strategic nuclear deterrent on time would have devastating impacts on deterrence and strategic stability. Accordingly, the committee directs the CNO to pay particular attention in producing the report to the Navy's ability to fully meet STRATCOM requirements. (Pages 172-174)

Section 1023 states:

SEC. 1023. REPEAL OF POLICY RELATING TO PROPULSION SYSTEMS OF ANY NEW CLASS OF MAJOR COMBATANT VESSELS OF THE STRIKE FORCES OF THE UNITED STATES NAVY.

Section 1012 of the National Defense Authorization Act for Fiscal Year 2008 (10 U.S.C. 7291 note) is repealed.

Regarding Section 1023, S.Rept. 113-44 states:

Repeal of policy relating to propulsion systems of any new class of major combatant vessels of the strike forces of the United States Navy (sec. 1023)

The committee recommends a provision that would repeal section 1012 of the National Defense Authorization Act for Fiscal Year 2008 (Public Law 110–181). That section requires that the Navy build any new class of major surface combatant and amphibious assault ship with an integrated nuclear power system, unless the Secretary of the Navy notifies the congressional defense committees that, as a result of a cost-benefit analysis, it would not be practical for the Navy to design the class of ships with an integrated nuclear power system.

As a matter of acquisition management laws and regulations, the Navy must conduct an analysis of alternatives (AoA) for all such vessels to determine the ship characteristics, including the required propulsion and power generation mechanism. The AoA outcome for each surface combatant and amphibious assault ship should be based on an independent and objective systems engineering and business case analysis. Within that analysis, the Navy can evaluate the advisability of using an integrated nuclear propulsion system to determine the best value for the Federal Government. (Pages 174-175)

S.Rept. 113-44 also states:

Ship Modernization, Operations and Sustainment Fund

Section 8105 of the Defense Appropriations Act for Fiscal Year 2013 (Public Law 113–6) established the Ship Modernization, Operations and Sustainment Fund, and appropriated more than $2.4 billion to the Fund. The Fund was intended to prevent the premature retirement of seven cruisers and two dock landing ships during fiscal years 2013 and 2014. This reflected a concern with the proposed retirement plan that the plan: (1) was disconnected from the defense strategy; (2) created future unaffordable shipbuilding requirements; and (3) would exacerbate force structure shortfalls that negatively impact the Department's ability to meet combatant commander (COCOM) requirements.

> The Report to Congress on the Annual Long-Range Plan for Construction of Naval Vessels for fiscal year 2014, date May 10, 2013, proposes to retire these cruisers and amphibious ships during fiscal year 2015, resulting in a fleet of 270 ships, the smallest fleet since 1917. The Navy is taking this action despite the fact that keeping these vessels operating until the end of 2014 will cost, according to the Navy, $931.1 million. The committee believes that the Navy should use the remaining resources in the Fund to sustain all of these ships. Available funds would permit the Navy to operate the ships during most of the period future- years defense program and would permit the Navy and Congress to continue evaluating options for modernizing and retaining these vessels until the end of their expected service lives. (Pages 30-31)

S.R ept. 113-44 also states:

> **LHA–8 design effort**
>
> The budget request included $155.3 million in PE [Program Element] 64567N for various ship design and research and development efforts, including $30.8 million for the next amphibious assault ship, LHA–8. Within the $30.8 million, $14.5 million is for LHA–8 ship design. Navy LHA–8 program development and design activities have involved two shipyards, among other contractors. The Navy intends to begin procurement funding for LHA–8 in fiscal year 2015.
>
> Repeated Navy shipbuilding programs have shown that failing to complete a ship's design before starting construction inevitably leads to cost growth and schedule delays. The committee believes that the Navy should invest more than it is currently planning to invest in maturing the design of LHA–8 before starting construction activities.
>
> Therefore, the committee recommends an increase of $20.0 million in PE 64567N for maturing the LHA–8 design. (Page 44-45)

FY2014 DOD Appropriations Act (H.R. 2397/S. 1429)

House

The House Appropriations Committee, in its report (H.Rept. 113-113 of June 17, 2013) on H.R. 2397, generally supported the Navy's requests for funding for the procurement of 8 new battle force ships while recommending increases and reductions to some of the requested amounts. (Pages 163-164) The report recommended:

- a $950 million increase to the Virginia class submarine program to "Fully fund the Virginia class submarine program;"
- a $96.1 million reduction to the program for performing mid-life nuclear refueling overhauls on existing Nimitz (CVN-68) class carriers for an "Asset due to prior year above threshold reprogramming;"
- a $38 million increase for the Afloat Forward Staging Base (AFSB) program for "Program shortfall;"
- a $23.4 million increase for the Moored Training Ship program for "Program shortfall," matching the amount sequestered from the program by the March 1, 2013, sequester; and

- a $7.6 million increase for the JHSV program for "Program shortfall," which is about one-third as much as the $21.4 million sequestered from the program by the March 1, 2013, sequester.

H.Rept. 113-113 states:

SHIPBUILDING

Despite assurances from the Navy that the shipbuilding account is as healthy as it has ever been, the Committee remains skeptical. The Navy's new stated required fleet size is 306 ships, down from the long stated but never achieved fleet size of 313 ships. The annual 30 year shipbuilding report provided by the Navy shows that the fleet will reach the required fleet size of 306 ships in the year 2037, a full quarter century from now. The Secretary of the Navy testified that the fleet size will reach the 300 ship mark by the end of the decade. While this is encouraging, the Committee is concerned the fleet size remains at 300 ships for a single year, then drops back under 300 until 2024. In fact, of the 30 years that comprise the plan, the fleet size is under 300 ships for 16 of these years. The fleet size is at the required 306 ships or above for only the final seven years (2037 to 2043) that are displayed in the report. The Committee believes a truly healthy shipbuilding program should reach and sustain the required fleet size sooner than the Navy is projecting.

Additionally, the Navy requested authority to incrementally fund a Virginia Class Submarine in fiscal year 2014. The Congress has authorized and appropriated funding for 18 fully funded Virginia Class Submarines in the years prior to fiscal year 2014. The Committee does not understand why funding requested for this particular submarine requires violating the Department of Defense's long standing full funding policy. The Committee is puzzled by Navy claims of billions of dollars in savings for the taxpayers as it is the Committee's understanding that these savings come from the fact that the program is conducting a multi- year procurement of ten submarines and not from the fact that one of the submarines is incrementally funded. Quality budget discipline, not funding gimmicks, is called for more than ever in these times of decreasing budgets and budget uncertainty.

Contributing to the poor health of the shipbuilding program is sequestration. As a direct result of sequestration, a minimum of seven destroyers and submarines from fiscal year 2013 and prior years are no longer fully funded. The cost to make the funding for these ships whole is just over $1,000,000,000. This funding will be required in the outyears to complete the construction of these ships. When this funding is added to the funding required to complete the incrementally funded submarine, the cost to complete these ships is in excess of $2,000,000,000. This represents funding that will not be available to purchase new equipment or increase readiness in future years, because it will have to pay the debt incurred by the Navy in years past. While sequestration is in statute, the Committee is extremely concerned about the Navy willfully adding to its outyear liabilities by incrementally funding a submarine. Therefore, the recommendation provides an additional $950,000,000 to the Virginia Class Submarine program to fully fund the program. Additionally, the Secretary of the Navy is directed to utilize the fiscal year 2015 funding currently reserved for the completion of the submarine to fully fund the ships and programs that were impacted by the sequestration reductions.

JOINT HIGH SPEED VESSEL

The Committee encourages the Secretary of the Navy to continue to explore missions and projects that leverage the flexibility of the Joint High Speed Vessel and explore the extension of the current mission envelope beyond in-theater transport, to include such capabilities as unmanned aerial systems and air surveillance.

SHIP DECOMMISSIONINGS

The Committee is extremely disappointed in the Navy's inaction with respect to the seven cruisers and two amphibious ships that were proposed for decommissioning in the fiscal year 2013 budget. Despite very clear direction from all four congressional defense committees to keep these ships in the fleet, the Navy has taken no steps that would indicate it is moving toward keeping the ships for the long term. The ships have significant life remaining and are less expensive to keep when compared to procuring new ships with similar capabilities. Last year the Congress provided sufficient funding for the operation and modernization of these ships through the end of fiscal year 2014 and that funding has gone largely untouched, indicating an unwillingness to commit to keep these ships in the fleet. The Committee fails to understand why the Navy would choose to decommission these ships when it is having such a difficult time maintaining the required fleet size through new procurements.

Therefore, the Committee recommendation rescinds all modernization funding from the fiscal year 2013 Ship Modernization, Operations, and Sustainment Fund and re-appropriates the funding in the Other Procurement, Navy, Weapons Procurement, Navy, and Research, Development, Test and Evaluation, Navy accounts. The Secretary of the Navy is directed to use this funding for the purpose of modernization of these seven cruisers and two amphibious ships and to retain them in the fleet. (Pages 165-166)

Senate

The Senate Appropriations Committee, in its report (S.Rept. 113-85 of August 1, 2013) on S. 1429, generally supported the Navy's requests for funding for the procurement of eight new battle force ships while recommending increases and reductions to some of the requested amounts. (Pages 100-101) The report recommended:

- a $27.313 million reduction to the CVN-78 aircraft carrier program for "Restoring acquisition accountability: Reduction in change orders" ($16.2 million) and "Maintaining program affordability: SEWIP [Surface Electronic Warfare Improvement Program] Block 3 excessive cost growth" ($11.113 million);
- a $22.1 million reduction to the program for performing mid-life nuclear refueling overhauls on existing Nimitz (CVN-68) class carriers for an "Improving funds management: CVN 72 requirement previously funded in Fiscal Year 2012 reprogramming";
- a $227 million increase to the Virginia class attack submarine program for "Maintain critical industrial base: Virginia class submarine";
- a $91 million net increase to the DDG-51 destroyer program, consisting of a $100 million increase for "Authorization adjustment: DDG–51" and a $9 million reduction for "Restoring acquisition accountability: Flight III Advance Planning early to need";
- a $55.3 million increase for the Afloat Forward Staging Base (AFSB) program for "Improving funds management: Transfer from NDSF [National Defense Sealift Fund], line 020, for full funding of ASFB #2 only, per Navy request";
- a $23.4 million increase for the Moored Training Ship program for "Improving funds management: Program shortfall, funds transferred per Navy request"; and

- a $7.6 million increase to the Joint High Speed Vessel (JHSV) program for "Improving funds management: JHSV program shortfall, funds transferred per Navy request."

S.Rept. 113-85 states:

Amphibious-class Warship Construction.—The Committee notes that the fiscal year 2014 budget request submitted by the Navy, once again, failed to present a plan to address the amphibious lift shortfall that exists today. In January 2009, the Navy and Marine Corps determined a minimum force of 33 ships is the limit of acceptable risk in meeting a 38-ship amphibious warship force requirement. As of now, there are 28 ships in the Navy's amphibious fleet, with an average of only 22 ships operationally available at any given time due to maintenance and overhaul schedules. This level of assumed risk with amphibious lift capability by the Department of the Navy deeply concerns the Committee. Of particular concern is the impact it has on the ability of Commanders to meet operations plans and crisis response requirements, particularly as instability in the Middle East continues and as the Department of Defense rebalances its global posture towards the Asia-Pacific region.

The ability to address the amphibious lift shortfall is exacerbated when the Navy funds only one amphibious class warship in the current 5-year Future Years Defense Program. This lack of commitment and funding by the Navy will not only have a negative impact on meeting future operations plans and crisis response requirements, but it will also have a negative industrial base impact and lead to additional cost growth in multiple shipbuilding programs. Therefore, the Committee directs the Secretary of the Navy to provide a more responsible amphibious warship acquisition plan to Congress with the fiscal year 2015 budget submission.

DDG–51 Destroyer.—The Committee understands that the DDG–51 program has a $304,000,000 shortfall due to prior year sequestration reductions, and recommends an additional $100,000,000 for the DDG–51 to allow the Navy to award the tenth DDG–51 under the current multi-year procurement contract, as previously authorized and appropriated. The Committee understands that its fiscal year 2014 shipbuilding recommendations create an outyear asset for the Navy to apply to shortfalls.

Joint High Speed Vessel.—The Committee recommends that the Navy continue to explore missions and projects that leverage the flexibility of the Joint High Speed Vessel [JHSV] and extend the mission envelope beyond in-theatre transport, including considering the addition of an unmanned aerial system and air surveillance capability to the JHSV. (Pages 101-102)

S.R ept. 113-85 also states:

SHIP MODERNIZATION, OPERATIONS AND SUSTAINMENT FUND

With the fiscal year 2014 budget submission, the Navy again proposes to prematurely retire seven Ticonderoga-class guided missile cruisers and two amphibious dock landing ships that have a combined remaining service life of over 100 years. The Committee notes that this proposal was rejected by the Congress in the Fiscal Year 2013 National Defense Authorization Act and the Fiscal Year 2013 Department of Defense Appropriations Act; and that Congress provided significant funds to man, operate, sustain and modernize these ships. As previously expressed in Senate Report 112–196, the Committee is concerned with this proposed elimination of force structure and believes it is disconnected from the strategic shift to the Asia-Pacific, creates future unaffordable shipbuilding requirements, and exacerbates force structure shortfalls that negatively impact the Department's ability to meet Combatant Command requirements.

The Committee notes that some key assumptions that led the Navy to propose prematurely retiring these ships have changed. This includes the material condition of at least one ship being superior to what the Navy assumed, as well as the scope and cost of modernization efforts required for these platforms to maintain their operational relevance for the balance of their service lives. The Committee believes that further adjustments to projected modernization efforts could be made, resulting in cost savings while retaining valuable operational capability in the near-term.

Therefore, the Committee again recommends denying these proposed premature retirements and retaining this force structure in its entirety. The Committee recommends $2,422,400,000, to man, operate, sustain, upgrade and modernize only CG–63, CG–64, CG–65, CG–66, CG– 68, CG–69, CG–73, LSD–41 and LSD–46 in the Ship Modernization, Operations and Sustainment Fund, as specified elsewhere in this act. Recognizing the time required to plan and execute shipyard availabilities and modernization periods, the Committee makes these funds available until September 30, 2021. However, the Committee also believes that upgrades to these ships have been delayed for too long, and therefore directs the Secretary of the Navy to upgrade at least one of the above listed Ticonderoga-class cruisers starting in fiscal year 2014. The Committee believes that this recommendation provides the fiscal relief required by the Navy to maintain this critical force structure and allows the Navy sufficient time to budget for this force structure in future budget submissions. (Page 10)

Funding for the Ship Modernization, Operations and Sustainment Fund is provided by **Section 8102** of S. 1429 as reported by the committee. The text of Section 8102 is as follows:

Sec. 8102. (a) Of the funds previously appropriated for the `Ship Modernization, Operations and Sustainment Fund', $2,098,000,000 is hereby rescinded;

(b) There is appropriated $2,422,400,000 for the `Ship Modernization, Operations and Sustainment Fund', to remain available until September 30, 2021: Provided, That the Secretary of the Navy shall transfer funds from the `Ship Modernization, Operations and Sustainment Fund' to appropriations for military personnel; operation and maintenance; research, development, test and evaluation; and procurement, only for the purposes of manning, operating, sustaining, equipping and modernizing the Ticonderoga-class guided missile cruisers CG-63, CG-64, CG-65, CG-66, CG-68, CG-69, CG-73, and the Whidbey Island-class dock landing ships LSD-41 and LSD-46: Provided further, That funds transferred shall be merged with and be available for the same purposes and for the same time period as the appropriation to which they are transferred: Provided further, That the transfer authority provided herein shall be in addition to any other transfer authority available to the Department of Defense: Provided further, That the Secretary of the Navy shall, not less than 30 days prior to making any transfer from the `Ship Modernization, Operations and Sustainment Fund', notify the congressional defense committees in writing of the details of such transfer: Provided further, That the Secretary of the Navy shall transfer and obligate funds from the `Ship Modernization, Operations and Sustainment Fund' for modernization of not less than one Ticonderoga-class guided missile cruiser as detailed above in fiscal year 2014.

Concerning the Navy's research and development account, S.Rept. 113-85 states:

LHA 8 Amphibious Assault Ship.—The fiscal year 2014 budget request includes $19,967,000 [in research and development funding] for LHA 8 amphibious assault ship preliminary design efforts. The Committee is aware that the Department of the Navy plans to reintroduce a well deck and optimize the aviation capability of LHA 8, which is planned for initial procurement funding in fiscal year 2015. As described in the Senate report accompanying S. 1197, the National Defense Act for Fiscal Year 2014, as reported, previous Navy efforts to start ship construction prior to completing a ship's design inevitably led to cost growth and schedule delays. Considering growing fiscal pressure on the national defense budget and increasing amphibious assault ship demands from combatant commanders for contingency operations, theater security cooperation, humanitarian assistance, and conventional deterrence missions, the Committee believes it is essential that LHA 8 be introduced in the most cost-effective manner. Therefore, the Committee includes an additional $50,000,000 for LHA 8 advance planning and design and directs the Department of the Navy to work with industry to identify affordability and producibility strategies that will lead to more efficient construction of a large deck amphibious assault ship. (Pages 161-162)

Appendix A. 2012 Testimony on Size of Navy Needed to Fully Meet COCOM Requests

This appendix presents three instances in 2012 when Navy officials testified that a Navy of more than 500 ships would be required to fully meet combatant commander COCOM requests for Navy forces.

March 22, 2012, Hearing

At a March 22, 2012, hearing on Navy readiness before the Readiness subcommittee of the House Armed Services Committee, the following exchange occurred:

> REPRESENTATIVE J. RANDY FORBES:
>
> We have a lot of requests for our combatant commanders—of the validated requests that come from combatant commanders. How many ships would it take in our navy, based on your estimation, to meet all of the validated requests from our commanders, combatant commanders?
>
> VICE ADMIRAL WILLIAM R. BURKE, DEPUTY CHIEF OF NAVAL OPERATIONS FOR FLEET READINESS AND LOGISTICS:
>
> Let me—give me just a minute on that, sir.
>
> FORBES:
>
> Please. And if you'd like, on any of these questions, if you'd rather take them for the record and get back I'm OK with that, too.
>
> BURKE:
>
> I'm—no, I'm happy to answer the question. I just want to make sure I elaborate a little to make sure we get—get the point right.
>
> FORBES:
>
> Please.
>
> BURKE:
>
> The—the combatant commander requests come in to the—to the services, and then the— there's a—a very high number of requirements from the services, or from the—the combatant commanders which are then prioritized and adjudicated by the joint staff.
>
> Essentially, a way to adjudicate supply—a lesser supply and a greater demand. So—so those—of those requests that come in, some are determined to be more valid than

others, if you will. But to get to your exact question, of those requests that come in from the combatant commanders, if we ...

(CROSSTALK)

FORBES:

Admiral, could—could I just—on the nomenclature, just make sure I'm right, too. As they come in, one of the first weed-out processes is we determine whether they're validated or not. In other words, we go through and make sure they're legal, they don't have the other asset somewhere. And—and then we stamp them as validated.

And then like you said, they go through a process where we then look at the resources we have and allocate what we can. And we adjudicate which ones we can give and which ones we can't. So I want the top number. The—the ones that we have validated and said, "Yes, this is legal, it's a proper request."

Of those combatant commander requests, approximately how many ships would it take us to be able to meet those if we had them?

BURKE:

It would take a navy of over 500 ships to meet the combatant commander requests. And, of course, it would take a similar increase in the aircraft and—and other parts of the—of the Navy, as well, to meet the combatant commander requests.[55]

March 29, 2012, Hearing

At a March 29, 2012, hearing on Navy shipbuilding programs before the Seapower and Projection Forces subcommittee of the House Armed Services Committee, the following exchange occurred:

REPRESENTATIVE DUNCAN D. HUNTER:

If you were to build the amphibs [i.e., amphibious ships] where would you prioritize? I mean, where would you take money out of to be able to get the Marine Corps to where they need to be?

VICE ADMIRAL JOHN TERENCE BLAKE, DEPUTY CHIEF OF NAVAL OPERATIONS FOR INTEGRATION OF CAPABILITIES AND RESOURCES:

Here's the issue we deal with: I don't have the luxury of dealing with any single issue in isolation; I have to deal with it across the entire...

HUNTER:

Well, we can. That's why I'm asking.

BLAKE:

Well, we have to deal with it, though, across the entire portfolio.

HUNTER:

Sure.

BLAKE:

And so what we have to do is we have to balance the requirement for amphibs, the requirement for surface combatants, the requirement for the carriers, the submarines—every category of ships that we have. And so when we do that we then have to say, all right, as we balance across that where are we going to be able to assume more risk? And that's how we— that's how we end up where we are.

HUNTER:

So you're saying there is less risk but still risk in the Marine Corps being short on amphibs than there are in the other—the rest of the picture?

BLAKE:

No. I'm saying that we have assumed risk in all areas. The best example I can give you: It was only a short time ago, if we tried to fill all the COCOM needs we said the number was around 400 ships we'd need in the fleet. Today—and we see no abatement in that commitment or the...

HUNTER:

No (inaudible) signal. BLAKE:

Today we look at it and we see that we would—if we wanted to hit 100 percent of all the COCOM requirements we'd need in excess of 500 ships. So what we end up having to do is we go through the—the global management process and we look at it and we say, here are our highest priorities, these are how we are going to address them, and then we—we have those units available and we push that...

HUNTER:

I understand.

I'm going to yield back in just one second.

So I would take from your statement, then, that you did go through a prioritization process and the amphibs are not at the top of that list. And second, when you say that you assume risk all the way around I would argue that when you do your risk assessment and you prioritize your needs the fact that the COCOMs wanted more ships and needed more ships due to the international environment and where we find ourselves with the world today, going down is probably – it's going the wrong way.

We all know that, but I—I would—I would argue that your prioritization—I would like to see that, if you don't mind, the—the way that you analyzed this and the—and the way that you said, hey, we're going to—we're going to keep them there to make sure that we have this over here. That's all I'm asking for.

BLAKE:

OK. When we put it together we do it across the entire spectrum; we don't—and by that I mean, as we look at the entire requirement we say, this is what we need to do in order to be able to meet the COCOM demand signal.[56]

April 19, 2012, Hearing

At an April 19, 2012, hearing on Navy shipbuilding programs before the Seapower subcommittee of the Senate Armed Services Committee, the following exchange occurred:

SENATOR ROGER WICKER:

Admiral Burke and General Mills, from an operational perspective, the Navy budget cost for decreases in large amphibious ships among other categories, in my opening statement, I mentioned that the requests from combatant commanders for amphibious ships has increased over 80 percent in the last five years—a very dramatic number. What is the reason for that and what will be the impact if these requests are—are not met?

VICE ADMIRAL WILLIAM R. BURKE, DEPUTY CHIEF OF NAVAL OPERATIONS FOR FLEET READINESS AND LOGISTICS:

Senator, thanks for the question. You're right, the COCOM demand signal has gone up significantly to the point where if we were to meet their—all their requirements, it would take a Navy of greater than 500 ships.

So, I certainly am not here to begrudge the COCOM demand signal because they have challenges that they're trying to deal with. But—but we can't meet—meet all their demands. So there is a process in the—in the Pentagon run by the joint staff called the Global Force Management process by which they take in the COCOM requirements and adjudicate that along with the forces we have to come to a reasonable allocation of—of force. And so that's the—that's a process we're dealing with today. We've been using that process for a number of years, and I would expect we will continue to use that process in the future to—to bridge the gap between supply and demand.[57]

APPENDIX B. COMPARING PAST SHIP FORCE LEVELS TO CURRENT OR POTENTIAL FUTURE SHIP FORCE LEVELS

In assessing the appropriateness of the current or potential future number of ships in the Navy, observers sometimes compare that number to historical figures for total Navy fleet size. Historical figures for total fleet size, however, can be a problematic yardstick for assessing the appropriateness of the current or potential future number of ships in the Navy, particularly if the historical figures are more than a few years old, because

- the missions to be performed by the Navy, the mix of ships that make up the Navy, and the technologies that are available to Navy ships for performing missions all change over time; and
- the number of ships in the fleet in an earlier year might itself have been inappropriate (i.e., not enough or more than enough) for the meeting the Navy's mission requirements in that year.

Regarding the first bullet point above, the Navy, for example, reached a late-Cold War peak of 568 battle force ships at the end of FY1987,[58] and as of November 8, 2013, included a total of 285 battle force ships. The FY1987 fleet, however, was intended to meet a set of mission requirements that focused on countering Soviet naval forces at sea during a potential multi-theater NATO-Warsaw Pact conflict, while the November 2013 fleet is intended to meet a considerably different set of mission requirements centered on influencing events ashore by countering both land- and sea-based military forces of potential regional threats other than Russia, including improved Chinese military forces and non-state terrorist organizations. In addition, the Navy of FY1987 differed substantially from the November 2013 fleet in areas such as profusion of precision-guided air-delivered weapons, numbers of Tomahawk-capable ships, and the sophistication of C4ISR systems and networking capabilities.[59]

In coming years, Navy missions may shift again, and the capabilities of Navy ships will likely have changed further by that time due to developments such as more comprehensive implementation of networking technology, increased use of ship-based unmanned vehicles, and the potential fielding of new types of weapons such as lasers or electromagnetic rail guns.

The 568-ship fleet of FY1987 may or may not have been capable of performing its stated missions; the 285-ship fleet of November 2013 may or may not be capable of performing its stated missions; and a fleet years from now with a certain number of ships may or may not be capable of performing its stated missions. Given changes over time in mission requirements, ship mixes, and technologies, however, these three issues are to a substantial degree independent of one another.

For similar reasons, trends over time in the total number of ships in the Navy are not necessarily a reliable indicator of the direction of change in the fleet's ability to perform its stated missions. An increasing number of ships in the fleet might not necessarily mean that the fleet's ability to perform its stated missions is increasing, because the fleet's mission requirements might be increasing more rapidly than ship numbers and average ship capability. Similarly, a decreasing number of ships in the fleet might not necessarily mean that the fleet's ability to perform stated missions is decreasing, because the fleet's mission requirements might be declining more rapidly than numbers of ships, or because average ship capability and the percentage of time that ships are in deployed locations might be increasing quickly enough to more than offset reductions in total ship numbers.

Regarding the second of the two bullet points above, it can be noted that comparisons of the size of the fleet today with the size of the fleet in earlier years rarely appear to consider whether the fleet was appropriately sized in those earlier years (and therefore potentially suitable as a yardstick of comparison), even though it is quite possible that the fleet in those earlier years might not have been appropriately sized, and even though there might have been differences of opinion among observers at that time regarding that question. Just as it might not be prudent for observers years from now to tacitly assume that the 285-ship of November 2013 was appropriately sized for meeting the mission requirements of 2013, even though there currently are differences of opinion among observers on that question (as reflected, for example, in **Table 7**) simply because a figure of 285 ships appears in the historical records for 2013, so, too, might it not be prudent for observers today to tacitly assume that the number of ships of the Navy in an earlier year was appropriate for meeting the Navy's mission requirements that year, even though there might have been differences of opinion among observers at that time regarding that question, simply because the size of the Navy in that year appears in a table like **Table D-1**.

Previous Navy force structure plans, such as those shown in **Table 1**, might provide some insight into the potential adequacy of a proposed new force-structure plan, but changes over time in mission requirements, technologies available to ships for performing missions, and other force- planning factors, as well as the possibility that earlier force-structure plans might not have been appropriate for meeting the mission demands of their times, suggest that some caution should be applied in using past force structure plans for this purpose, particularly if those past force structure plans are more than a few years old. The Reagan-era plan for a 600-ship Navy, for example, was designed for a Cold War set of missions focusing on countering Soviet naval forces at sea, which is not an appropriate basis for planning the Navy today, and there was considerable debate during those years as to the appropriateness of the 600-ship goal.[60]

APPENDIX C. INDEPENDENT PANEL ASSESSMENT OF 2010 QDR

The law that requires DOD to perform QDRs once every four years (10 U.S.C. 118) states that the results of each QDR shall be assessed by an independent panel. The report of the independent panel that assessed the 2010 QDR was released on July 29, 2010. The independent panel's report recommended a Navy of 346 ships, including 11 aircraft carriers and 55 attack submarines.[61] The report stated the following, among other things:

- "The QDR should reflect current commitments, but it must also plan effectively for potential threats that could arise over the next 20 years.... we believe the 2010 QDR did not accord sufficient priority to the need to counter anti-access challenges, strengthen homeland defense (including our defense against cyber threats), and conduct post-conflict stabilization missions." (Page 54)

- "In this remarkable period of change, global security will still depend upon an American presence capable of unimpeded access to all international areas of the Pacific region. In an environment of 'anti-access strategies,' and assertions to create unique 'economic and security zones of influence,' America's rightful and historic presence will be critical. To preserve our interests, the United States will need to retain the ability to transit freely the areas of the Western Pacific for security and economic reasons. Our allies also depend on us to be fully present in the Asia-Pacific as a promoter of stability and to ensure the free flow of commerce. A robust U.S. force structure, largely rooted in maritime strategy but including other necessary capabilities, will be essential." (Page 51)
- "The United States will need agile forces capable of operating against the full range of potential contingencies. However, the need to deal with irregular and hybrid threats will tend to drive the size and shape of ground forces for years to come, whereas the need to continue to be fully present in Asia and the Pacific and other areas of interest will do the same for naval and air forces." (Page 55)
- "The force structure in the Asia-Pacific needs to be increased. In order to preserve U.S. interests, the United States will need to retain the ability to transit freely the areas of the Western Pacific for security and economic reasons. The United States must be fully present in the Asia-Pacific region to protect American lives and territory, ensure the free flow of commerce, maintain stability, and defend our allies in the region. A robust U.S. force structure, one that is largely rooted in maritime strategy and includes other necessary capabilities, will be essential." (Page 66)
- "Force structure must be strengthened in a number of areas to address the need to counter anti-access challenges, strengthen homeland defense (including defense against cyber threats), and conduct post-conflict stabilization missions: First, as a Pacific power, the U.S. presence in Asia has underwritten the regional stability that has enabled India and China to emerge as rising economic powers. The United States should plan on continuing that role for the indefinite future. The Panel remains concerned that the QDR force structure may not be sufficient to assure others that the United States can meet its treaty commitments in the face of China's increased military capabilities. Therefore, we recommend an increased priority on defeating anti-access and area-denial threats. This will involve acquiring new capabilities, and, as Secretary Gates has urged, developing innovative concepts for their use. Specifically, we believe the United States must fully fund the modernization of its surface fleet. We also believe the United States must be able to deny an adversary sanctuary by providing persistent surveillance, tracking, and rapid engagement with high-volume precision strike. That is why the Panel supports an increase in investment in long-range strike systems and their associated sensors. In addition, U.S. forces must develop and demonstrate the ability to operate in an information-denied environment." (Pages 59-60)
- "To compete effectively, the U.S. military must continue to develop new conceptual approaches to dealing with operational challenges, like the Capstone Concept for Joint Operations (CCJO). The Navy and Air Force's effort to develop an Air-Sea Battle concept is one example of an approach to deal with the growing anti-access challenge. It will be necessary to invest in modernized capabilities to make this happen. The Chief of Naval Operations and Chief of Staff of the Air Force deserve

support in this effort, and the Panel recommends the other military services be brought into the concept when appropriate." (Page 51; a similar passage appears on page 67)

Table C-1. Comparison of Navy's 306-ship goal, Navy Plan from 1993 BUR, and Navy Plan from 2010 QDR Review Panel

Ship Type	Navy's 306-ship goal of March 2012	Bottom-Up Review (BUR) (1993)	2010 QDR Independent Review Panel (July 2010)
SSBNs	12-14	18 (SSBN force was later reduced to 14 as a result of the 1994 Nuclear Posture Review)	14
SSGNs	0-4	0 (SSGN program did not yet exist)	4
SSNs	~48	45 to 55 (55 in FY99, with a long-term goal of about 45)	55
Aircraft carriers	11 active	11 active + 1 operational/reserve	11 active
Surface combatants	~145	124 (114 active + 10 frigates in Naval Reserve Force; a total of 110-116 active ships was also cited)	n/a
Cruisers and destroyers	*~90*	*n/a*	*n/a*
Frigates	*0* *(to be replaced by LCSs)*	*n/a*	*n/a*
LCSs	*~55*	*0* *(LCS program did not exist)*	*n/a*
Amphibious ships	~32 (30 operational ships needed to lift 2.0 MEBs)	41 (Enough to lift 2.5 MEBs)	n/a
Dedicated mine warfare ships	0 (to be replaced by LCSs)	26 (LCS program did not exist)	n/a
CLF ships	~29	43	n/a
Support ships	~33	22	n/a
TOTAL ships	**~306**	**346** **(numbers above add to 331-341)**[a]	**346**

Source: Table prepared by CRS. ***Sources for 1993 Bottom-Up Review:*** Department of Defense, *Report on the Bottom-Up Review*, October 1993, Figure 7 on page 28; Department of the Navy, *Highlights of the FY 1995 Department of the Navy Budget*, February 1994, p. 1; Department of the Navy, *Force 2001, A Program Guide to the U.S. Navy*, 1994 edition, p. 15; Statement of VADM T. Joseph Lopez, U.S. Navy, Deputy Chief of Naval Operations (Resources, Warfare Requirements & Assessments), Testimony to the Military Forces and Personnel Subcommittee of the House Armed Services Committee, March 22, 1994, pp. 2-5. ***Source for independent panel report:*** Stephen J. Hadley and William J. Perry, co-chairmen, et al., *The QDR in Perspective: Meeting America's National Security Needs In the 21st Century, The Final Report of the Quadrennial Defense Review Independent Panel*, Washington, 2010, Figure 3-2 on pages 58-59.

Notes: **n/a** is not addressed in the report. **SSBN** is nuclear-powered ballistic missile submarine; **SSGN** is nuclear-powered cruise missile and special operations forces submarine; **SSN** is nuclear-powered attack submarine; **LCS** is Littoral Combat Ship; **MPF(F)** is Maritime Prepositioning Force (Future) ship; **CLF** is combat logistics force (i.e., resupply) ship; **MEB** is Marine Expeditionary Brigade.

a. The Navy testified in 1994 that the planned number was adjusted from 346 to 330 to reflect reductions in numbers of tenders and early retirements of some older amphibious ships.

In recommending a Navy of 346 ships, the independent panel's report cited the 1993 Bottom-Up Review (BUR) of U.S. defense plans and policies. **Table C-1** compares the Navy's 306-ship goal of March 2012 to the 346-ship Navy recommended in the 1993 BUR (as detailed partly in subsequent Navy testimony and publications) and the ship force levels recommended in the independent panel report.

In a letter dated August 11, 2010, Secretary of Defense Robert Gates provided his comments on the independent panel's report. The letter stated in part:

> I completely agree with the Panel that a strong navy is essential; however, I disagree with the Panel's recommendation that DoD should establish the 1993 Bottom Up Review's (BUR's) fleet of 346 ships as the objective target. That number was a simple projection of the then- planned size of [the] Navy in FY 1999, not a reflection of 21st century, steady-state requirements. The fleet described in the 2010 QDR report, with its overall target of 313 to 321 ships, has roughly the same number of aircraft carriers, nuclear-powered attack submarines, surface combatants, mine warfare vessels, and amphibious ships as the larger BUR fleet. The main difference between the two fleets is in the numbers of combat logistics, mobile logistics, and support ships. Although it is true that the 2010 fleet includes fewer of these ships, they are all now more efficiently manned and operated by the Military Sealift Command and meet all of DoD's requirements....
>
> I agree with the Panel's general conclusion that DoD ought to enhance its overall posture and capabilities in the Asia-Pacific region. As I outlined in my speech at the Naval War College in April 2009, "to carry out the missions we may face in the future... we will need numbers, speed, and the ability to operate in shallow waters." So as the Air-Sea battle concept development reaches maturation, and as DoD's review of global defense posture continues, I will be looking for ways to meet plausible security threats while emphasizing sustained forward presence – particularly in the Pacific.[62]

Appendix D. Size of the Navy and Navy Shipbuilding Rate

Size of the Navy

Table D-1 shows the size of the Navy in terms of total number of ships since FY1948; the numbers shown in the table reflect changes over time in the rules specifying which ships count toward the total. Differing counting rules result in differing totals, and for certain years, figures reflecting more than one set of counting rules are available. Figures in the table for FY1978 and subsequent years reflect the battle force ships counting method, which is the set of counting rules established in the early 1980s for public policy discussions of the size of the Navy.

Table D-1.Total Number of Ships in the Navy Since FY1948

FY[a]	Number	FY[a]	Number	FY[a]	Number
1948	737	1970	769	1992	466
1949	690	1971	702	1993	435
1950	634	1972	654	1994	391
1951	980	1973	584	1995	373
1952	1,097	1974	512	1996	356
1953	1,122	1975	496	1997	354
1954	1,113	1976	476	1998	333
1955	1,030	1977	464	1999	317
1956	973	1978	468	2000	318
1957	967	1979	471	2001	316
1958	890	1980	477	2002	313
1959	860	1981	490	2003	297
1960	812	1982	513	2004	291
1961	897	1983	514	2005	282
1962	959	1984	524	2006	281
1963	916	1985	541	2007	279
1964	917	1986	556	2008	282
1965	936	1987	568	2009	285
1966	947	1988	565	2010	288
1967	973	1989	566	2011	284
1968	976	1990	547	2012	287
1969	926	1991	526	2013	

Source: Compiled by CRS using U.S. Navy data. Numbers shown reflect changes over time in the rules specifying which ships count toward the total. Figures for FY1978 and subsequent years reflect the battle force ships counting method, which is the set of counting rules established in the early 1980s for public policy discussions of the size of the Navy.

a. Data for earlier years in the table may be for the end of the calendar year (or for some other point during the year), rather than for the end of the fiscal year.

As shown in the table, the total number of battle force ships in the Navy reached a late-Cold War peak of 568 at the end of FY1987 and began declining thereafter.[63] The Navy fell below 300 battle force ships in August 2003 and included 285 battle force ships as of November 8, 2013.

As discussed in **Appendix B**, historical figures for total fleet size might not be a reliable yardstick for assessing the appropriateness of proposals for the future size and structure of the Navy, particularly if the historical figures are more than a few years old, because the missions to be performed by the Navy, the mix of ships that make up the Navy, and the technologies that are available to Navy ships for performing missions all change over time, and because the number of ships in the fleet in an earlier year might itself have been inappropriate (i.e., not enough or more than enough) for the meeting the Navy's mission requirements in that year.

For similar reasons, trends over time in the total number of ships in the Navy are not necessarily a reliable indicator of the direction of change in the fleet's ability to perform its stated missions. An increasing number of ships in the fleet might not necessarily mean that the fleet's ability to perform its stated missions is increasing, because the fleet's mission requirements might be increasing more rapidly than ship numbers and average ship capability. Similarly, a decreasing number of ships in the fleet might not necessarily mean that the fleet's ability to perform stated missions is decreasing, because the fleet's mission requirements might be declining more rapidly than numbers of ships, or because average ship capability and the percentage of time that ships are in deployed locations might be increasing quickly enough to more than offset reductions in total ship numbers.

Shipbuilding Rate

Table D-2 shows past (FY1982-FY2013) and requested (FY2014-FY2018) rates of Navy ship procurement.

Table D-2. Battle Force Ships Procured or Requested, FY1982-FY2018 (Procured FY1982-FY2013; requested FY2014-FY2018*)*

82	**83**	**84**	**85**	**86**	**87**	**88**	**89**	**90**	**91**	**92**	**93**	**94**	**95**	**96**	**97**	**98**	**99**	**00**
17	14	16	19	20	17	15	19	15	11	11	7	4	4	5	4	5	5	6
01	**02**	**03**	**04**	**05**	**06**	**07**	**08**	**09**	**10**	**11**	**12**	**13**	***14***	***15***	***16***	***17***	***18***	
6	6	5	7	8	4[a]	5[a]	3[a]	8	7	10	11[b]	11[c]	*8*	*8*	*7*	*9*	*9*	

Source: CRS compilation based on Navy budget data and examination of defense authorization and appropriation committee and conference reports for each fiscal year. The table excludes non-battle force ships that do not count toward the 306-ship goal, such as certain sealift and prepositioning ships operated by the Military Sealift Command and oceanographic ships operated by agencies such as the National Oceanic and Atmospheric Administration (NOAA).

a. The totals shown for FY2006, FY2007, and FY2008, reflect the cancellation two LCSs funded in FY2006, another two LCSs funded in FY2007, and an LCS funded in FY2008.

b. The total shown for FY2012 includes two JHSVs—one that was included in the Navy's FY2012 budget submission, and one that was included in the Army's FY2012 budget submission. Until FY2012, JHSVs were being procured by both the Navy and the Army. The Army was to procure its fifth and final JHSV in FY2012, and this ship was included in the Army's FY2012 budget submission. In May 2011, the Navy and Army signed a Memorandum of Agreement (MOA) transferring the Army's JHSVs to the Navy. In the FY2012 DOD Appropriations Act (Division A of H.R. 2055/P.L. 112-74 of December 23, 2011), the JHSV that was in the Army's FY2012 budget submission was funded through the Shipbuilding and Conversion, Navy (SCN) appropriation account, along with the JHSV that the Navy had included in its FY0212 budget submission. The four JHSVs that were procured through the Army's budget prior to FY2012, however, are *not* included in the annual totals shown in this table.

c. Figure shown does not reflect potential quantity reduction resulting from March 1, 2013, sequester on FY2013 funding.

End Notes

[1] Department of the Navy, *Report to Congress [on] Navy Combatant Vessel Force Structure Requirement*, January 2013, 3 pp. The cover letters for the report were dated January 31, 2013.

[2] For more on this document, see CRS Report R42146, *In Brief: Assessing the January 2012 Defense Strategic Guidance (DSG)*, by Catherine Dale and Pat Towell.

[3] *Report to Congress on the Annual Long-Range Plan for Construction of Naval Vessels for FY2014*, May 2013, p. 4.

[4] For further discussion of this issue, see CRS Report R41129, *Navy Ohio Replacement (SSBN[X]) Ballistic Missile Submarine Program: Background and Issues for Congress*, by Ronald O'Rourke.

[5] *Department of Defense Report on the Joint Committee Sequestration for Fiscal Year 2013*, June 2013, pp. 36, 36A, 37, and 37A (pdf pages 85-88 of 438).

[6] Transcript of hearing. Stackley testified that the sequester created an approximately $560 million funding shortfall in the DDG-51 program (as mentioned earlier DOD's June 2013 report provided a final figure of $541.2 million), and that the Navy was able to identify about $260 million in funding offsets to apply to the shortfall, leaving a shortfall of about $300 million.

[7] Transcript of hearing.

[8] Statement of Admiral Jonathan Greenert, U.S. Navy, Chief of Naval Operations, Before the Senate Armed Services Committee on the Impact of Sequestration on the National Defense, November 7, 2013, pp. 11-12.

[9] Statement of Hon. Sean J. Stackley, Assistant Secretary of the Navy (Research, Development and Acquisition), and Vice Admiral Allen G. Myers, USN, Deputy Chief of Naval Operations, Integration of Capabilities and Resources, and Lieutenant General Glenn M. Walters, USMC, Deputy Commandant for Programs and Resources, Before the Tactical Air and Land Forces Subcommittee of the House Armed Services Committee on Impacts of a Continuing Resolution and Sequestration on Department of the Navy Acquisition, Programming & Industrial Base, October 23, 2013, pp. 2-3.

[10] Statement of Admiral Jonathan Greenert, U.S. Navy, Chief of Naval Operations, Before the House Armed Services Committee on Planning for Sequestration in FY 2014 and Perspectives of the Military Services on the Strategic Choices and Management Review, September 18, 2013, pp. 2-3

[11] Transcript of remarks by Admiral Jonathan Greenert, Chief of Naval Operations, at the American Enterprise Institute (AEI), September 5, 2013, on American Military Strategy In A Time of Declining Budgets, provided to CRS by Navy Office of Legislative Affairs, September 6, 2013.

[12] Statement of Rear Admiral Thomas Rowden, Director, Surface Warfare (N96) and Rear Admiral Timothy Matthews, Director, Fleet Readiness (N43) on Ensuring Navy Surface Force Effectiveness With limited Maintenance Resources before the House Armed Services Committee Subcommittee on Seapower and Projection Forces and Subcommittee on Readiness, August 1, 2013, pp. 1-2.

[13] Department of Defense Press Briefing by Adm. Greenert in the Pentagon Briefing Room, July 19, 2013, accessed August 7, 2013, at http://www.defense.gov/transcripts/transcript.aspx?transcriptid=5278. The term "fiscal year" appears in brackets in the original.

[14] Statement of The Honorable Sean J. Stackley, Assistant Secretary of the Navy (Research, Development and Acquisition) and Vice Admiral Allen G. Myers, Deputy Chief of Naval Operations for Integration of Capabilities and Resources and Vice Admiral Kevin M. McCoy, Commander, Naval Sea Systems Command, Before the Subcommittee on Seapower of the Senate Armed Services Committee on Department of the Navy Shipbuilding Programs, May 8, 2013, p. 5.

[15] Transcript of hearing.

[16] Transcript of hearing.

[17] Statement of Admiral Jonathan Greenert, U.S. Navy, Chief of Naval Operations, Before the Senate Armed Services Committee on the Impact of Sequestration on the National Defense, November 7, 2013, pp. 12-14.

[18] Statement of Hon. Sean J. Stackley, Assistant Secretary of the Navy (Research, Development and Acquisition), and Vice Admiral Allen G. Myers, USN, Deputy Chief of Naval Operations, Integration of Capabilities and Resources, and Lieutenant General Glenn M. Walters, USMC, Deputy Commandant for Programs and Resources, Before the Tactical Air and Land Forces Subcommittee of the House Armed Services Committee on Impacts of a Continuing Resolution and Sequestration on Department of the Navy Acquisition, Programming & Industrial Base, October 23, 2013, pp. 1-2, 12.

[19] Statement of Hon. Sean J. Stackley, Assistant Secretary of the Navy (Research, Development and Acquisition), and Vice Admiral Allen G. Myers, USN, Deputy Chief of Naval Operations, Integration of Capabilities and Resources, and Lieutenant General Glenn M. Walters, USMC, Deputy Commandant for Programs and Resources, Before the Tactical Air and Land Forces Subcommittee of the House Armed Services Committee on Impacts of a Continuing Resolution and Sequestration on Department of the Navy Acquisition, Programming & Industrial Base, October 23, 2013, pp. 3-8.

[20] Statement of Admiral Jonathan Greenert, U.S. Navy, Chief of Naval Operations, Before the House Armed Services Committee on Planning for Sequestration in FY 2014 and Perspectives of the Military Services on the Strategic Choices and Management Review, September 18, 2013, pp. 11-13.

[21] John Kirby, "U.S. Navy Presence At Risk Under Sequester," Navy Live (http://navylive.dodlive.mil), September 14, 2013, accessed September 24, 2013, at http://navylive.dodlive.mil/2013/09/14/presence-at-risk-under-sequester/.

[22] Transcript of remarks by Admiral Jonathan Greenert, Chief of Naval Operations, at the American Enterprise Institute (AEI), September 5, 2013, on American Military Strategy In A Time of Declining Budgets, provided to CRS by Navy Office of Legislative Affairs, September 6, 2013.

[23] Statement of Rear Admiral Thomas Rowden, Director, Surface Warfare (N96) and Rear Admiral Timothy Matthews, Director, Fleet Readiness (N43) on Ensuring Navy Surface Force Effectiveness With limited

Maintenance Resources before the House Armed Services Committee Subcommittee on Seapower and Projection Forces and Subcommittee on Readiness, August 1, 2013, pp. 2-3.

[24] A maintenance availability is a Navy term for a maintenance action, such as an overhaul.

[25] Department of Defense Press Briefing by Adm. Greenert in the Pentagon Briefing Room, July 19, 2013, accessed August 7, 2013, at http://www.defense.gov/transcripts/transcript.aspx?transcriptid=5278.

[26] Department of Defense Press Briefing by Adm. Greenert in the Pentagon Briefing Room, July 19, 2013, accessed August 7, 2013, at http://www.defense.gov/transcripts/transcript.aspx?transcriptid=5278.

[27] Department of Defense Press Briefing by Adm. Greenert in the Pentagon Briefing Room, July 19, 2013, accessed August 7, 2013, at http://www.defense.gov/transcripts/transcript.aspx?transcriptid=5278.

[28] Transcript of hearing.

[29] Statement of Admiral Jonathan Greenert, U.S. Navy, Chief of Naval Operations, Before the Senate Armed Services Committee on the Impact of Sequestration on the National Defense, November 7, 2013, pp. 7-11.

[30] Statement of Admiral Jonathan Greenert, U.S. Navy, Chief of Naval Operations, Before the House Armed Services Committee on Planning for Sequestration in FY 2014 and Perspectives of the Military Services on the Strategic Choices and Management Review, September 18, 2013, pp. 6-10.

[31] Transcript of remarks by Admiral Jonathan Greenert, Chief of Naval Operations, at the American Enterprise Institute (AEI), September 5, 2013, on American Military Strategy In A Time of Declining Budgets, provided to CRS by Navy Office of Legislative Affairs, September 6, 2013.

[32] Transcript of hearing.

[33] Transcript of hearing.

[34] The other Navy witness at the hearing was Rear Admiral Timothy Matthews, the Director for Fleet Readiness.

[35] Transcript of hearing.

[36] Department of Defense Press Briefing by Secretary Hagel and Adm. Winnefeld from the Pentagon, July 31, 2013, accessed August 7, 2013, at http://www.defense.gov/transcripts/transcript.aspx?transcriptid=5280.

[37] Transcript of hearing.

[38] Statement of Admiral Jonathan Greenert, Chief of Naval Operations, before the Senate Armed Services Committee on the Impact of Sequestration, February 12, 2013, pp. 8-9. Although the Navy's written statement for the hearing is from Admiral Greenert, the Navy's testimony at the hearing was actually given by the Vice Chief of Naval Operations, Admiral Mark E. Ferguson III, who was substituting for Admiral Greenert at the hearing. The Navy presented similar testimony at a similar hearing before the House Armed Services Committee on February 13, 2013.

Similarly, on October 22, 2012, Admiral Mark Ferguson, the Vice Chief of Naval Operations, stated, "If you project out 10 years—remember that the budget control act talks about 10 years of [funding] reductions—now you start talking about a fleet reduced to about 230-235 ships." (As quoted in David Smalley, "Leaner Navy Looking at Future Technology, Fleet Size and Sequestration," *Navy News Service*, October 23, 2012, accessed October 25, 2012 at http://www.navy.mil/submit/display.asp?story_id=70311.) On January 8, 2013, Vice Admiral William Burke, the Deputy Chief of Naval Operations for Warfare Systems, stated, "As a result of sequestration, what I testified to last year was we'll have, if sequestration is fully implemented, we'll have a force of about 230 ships.... I'm not talking about what the aviation numbers are, but they'll go down by a similar amount—a similar amount being 20 percent." (As quoted in Dan Taylor, "Adm. Burke: Navy Needs To Fill $2 Billion Hole When OCO Goes Away," *Inside the Navy*, January 14, 2013. See also Michael Fabey, "U.S. Navy Prioritizes Ship Total Ownership Costs, Maintenance," *Aerospace Daily & Defense Report*, January 9, 2013: 3.)

[39] For more on the modernization of China's military (particularly naval) capabilities and its potential implications for required U.S. Navy capabilities, see CRS Report RL33153, *China Naval Modernization: Implications for U.S. Navy Capabilities—Background and Issues for Congress*, by Ronald O'Rourke.

[40] Department of Defense, *Sustaining U.S. Global Leadership: Priorities for 21st Century Defense*, January 2012, 8 pp. For more on this document, see CRS Report R42146, *In Brief: Assessing the January 2012 Defense Strategic Guidance (DSG)*, by Catherine Dale and Pat Towell. See also CRS Report R42448, *Pivot to the Pacific? The Obama Administration's "Rebalancing" Toward Asia*, coordinated by Mark E. Manyin.

[41] See, for example, Dan Blumenthal and Michael Mazza, "Asia Needs a Larger U.S. Defense Budget," *Wall Street Journal*, July 5, 2011; J. Randy Forbes, "Defence Cuts Imperil US Asia Role," *The Diplomat (*http://the-diplomat.com*)*, October 26, 2011. See also Andrew Krepinevich, "Panetta's Challenge," *Washington Post*, July 15, 2011: 15; Dean Cheng, *Sea Power and the Chinese State: China's Maritime Ambitions*, Heritage Foundation Backgrounder No. 2576, July 11, 2011, p. 10.

[42] For further discussion of these options, see CRS Report RS21338, *Navy Ship Deployments: New Approaches—Background and Issues for Congress*, by Ronald O'Rourke.

[43] For further discussion, see CRS Report RL33153, *China Naval Modernization: Implications for U.S. Navy Capabilities—Background and Issues for Congress*, by Ronald O'Rourke.

[44] Congressional Budget Office, *An Analysis of the Navy's Fiscal Year 2014 Shipbuilding Plan*, October 2013, Table 3 (page 13).

[45] *Report to Congress on the Annual Long-Range Plan for Construction of Naval Vessels for FY2014*, May 2013, pp. 11-12, 18-19.

[46] Vice Admiral Burke retired on May 20, 2013.

[47] Walter Pincus, "Budget Cuts Could Reshape Country's Ship Supply," *Washington Post*, May 2, 2013: 15.

[48] This paragraph and those that follow are adapted from Statement of Ronald O'Rourke, Specialist in Naval Affairs, Congressional Research Service, Before the House Armed Services Committee Subcommittee on Seapower and Projection Forces, on the Navy's FY2014 30-Year Shipbuilding Plan, October 23, 2013, pp. 1-4.

[49] See *Report to Congress on the Annual Long-Range Plan for Construction of Naval Vessels for FY2014*, May 2013, p. 18.

[50] Congressional Budget Office, *An Analysis of the Navy's Fiscal Year 2014 Shipbuilding Plan*, October 2013, Table 3 (page 13).

[51] The "one-third, one-third, one-third" terminology, though convenient, is not entirely accurate—the military departments' shares of the DOD budget, while more or less stable during this period, were not exactly one-third each: the average share for the Department of the Army was about 26%, the average share for the Department of the Navy (which includes both the Navy and Marine Corps) was about 32%, the average share for the Department of the Air Force was about 30%, and the average share for Defense-Wide (the fourth major category of DOD spending) was about 12%. Excluding the Defense-Wide category, which has grown over time, the shares for the three military departments of the remainder of DOD's budget during this period become about 29% for the Department of the Army, about 37% for the Department of the Navy, and about 34% for the Department of the Air Force.

[52] Since the Defense-Wide portion of the budget has grown from just a few percent in the 1950s and 1960s to about 15% in more recent years, including the Defense-Wide category of spending in the calculation can lead to military department shares of the budget in the 1950s and 1960s that are somewhat more elevated compared to those in more recent years, making it more complex to compare the military departments' shares across the entire period of time since the end of the World War II. For this reason, military department shares of the DOD budget cited in this statement are calculated after excluding the Defense-Wide category. The points made in this statement, however, can still made on the basis of a calculation that includes the Defense-Wide category.

[53] Defense Business Board briefing, "Reducing Overhead and Improving Business Operations, Initial Observations," July 22, 2010, slides 15, 5, and 6, posted online at http://www.govexec.com/pdfs/072210rb1.pdf. See also Defense Business Board, *Modernizing the Military Retirement System*, Report to the Secretary of Defense, Report FY11-05, posted online at http://dbb.defense.gov/Portals/35/Documents/Reports/2011/FY11- 5_Modernizing_The_Military_Retirement_System_2011-7.pdf; and Defense Business Board, *Corporate Downsizing Applications for DoD*, Report to the Secretary of Defense, Report FY11-08, posted online at http://dbb.defense.gov/Portals/35/Documents/Reports/2011/FY11-8_Corporate_Downsizing_Applications_for_DoD_2011-7.pdf.

[54] Statement of Admiral Jonathan Greenert, U.S. Navy, Chief of Naval Operations, Before the House Armed Services Committee on Planning for Sequestration in FY 2014 and Perspectives of the Military Services on the Strategic Choices and Management Review, September 18, 2013, pp. 6-10.

[55] Source: Transcript of hearing.

[56] Source: Transcript of hearing.

[57] Source: Transcript of hearing.

[58] Some publications have stated that the Navy reached a peak of 594 ships at the end of FY1987. This figure, however, is the total number of active ships in the fleet, which is not the same as the total number of battle force ships. The battle force ships figure is the number used in government discussions of the size of the Navy. In recent years, the total number of active ships has been larger than the total number of battle force ships. For example, the Naval History and Heritage Command (formerly the Naval Historical Center) states that as of November 16, 2001, the Navy included a total of 337 active ships, while the Navy states that as of November 19, 2001, the Navy included a total of 317 battle force ships. Comparing the total number of active ships in one year to the total number of battle force ships in another year is thus an apples-to-oranges comparison that in this case overstates the decline since FY1987 in the number of ships in the Navy. As a general rule to avoid potential statistical distortions, comparisons of the number of ships in the Navy over time should use, whenever possible, a single counting method.

[59] C4ISR stands for command and control, communications, computers, intelligence, surveillance, and reconnaissance.

[60] Navy force structure plans that predate those shown in **Table 1** include the Reagan-era 600-ship plan of the 1980s, the Base Force fleet of more than 400 ships planned during the final two years of the George H. W. Bush Administration, the 346-ship fleet from the Clinton Administration's 1993 Bottom-Up Review (or BUR, sometimes also called Base Force II), and the 310-ship fleet of the Clinton Administration's 1997 QDR. The table below summarizes some key features of these plans.

Features of Recent Navy Force Structure Plans

Plan	600-ship	Base Force	1993 BUR	1997 QDR
Total ships	~600	~450/416[a]	346	~305/310[b]
Attack submarines	100	80/~55[c]	45-55	50/55[d]
Aircraft carriers	15[e]	12	11+1[f]	11+1[f]
Surface combatants	242/228[g]	~150	~124	116
Amphibious ships	~75[h]	51[i]	41[i]	36[i]

Source: Prepared by CRS based on DOD and U.S. Navy data.

a. Commonly referred to as 450-ship plan, but called for decreasing to 416 ships by end of FY1999.

b. Original total of about 305 ships was increased to about 310 due to increase in number of attack submarines to 55 from 50.

c. Plan originally included 80 attack submarines, but this was later reduced to about 55.

d. Plan originally included 50 attack submarines but this was later increased to 55.

e. Plus one additional aircraft carrier in the service life extension program (SLEP).

f. Eleven active carriers plus one operational reserve carrier.

g. Plan originally included 242 surface combatants but this was later reduced to 228.

h. Number needed to lift assault echelons of one Marine Expeditionary Force (MEF) plus one Marine Expeditionary Brigade (MEB).

i. Number needed to lift assault echelons of 2.5 MEBs. Changing numbers needed to meet this goal reflect in part changes in the design and capabilities of amphibious ships.

[61] Stephen J. Hadley and William J. Perry, co-chairmen, et al, *The QDR in Perspective: Meeting America's National Security Needs In the 21st Century, The Final Report of the Quadrennial Defense Review Independent Panel*, Washington, 2010, Figure 3-2 on page 58.

[62] Letter dated August 11, 2010, from Secretary of Defense Robert Gates to the chairmen of the House and Senate Armed Services and Appropriations Committees, pp. 3 and 4. The ellipsis in the second paragraph appears in the letter.

[63] Some publications have stated that the Navy reached a peak of 594 ships at the end of FY1987. This figure, however, is the total number of active ships in the fleet, which is not the same as the total number of battle force ships. The battle force ships figure is the number used in government discussions of the size of the Navy. In recent years, the total number of active ships has been larger than the total number of battle force ships. For example, the Naval History and Heritage Command (formerly the Naval Historical Center) states that as of November 16, 2001, the Navy included a total of 337 active ships, while the Navy states that as of November 19, 2001, the Navy included a total of 317 battle force ships. Comparing the total number of active ships in one year to the total number of battle force ships in another year is thus an apples-to-oranges comparison that in this case overstates the decline since FY1987 in the number of ships in the Navy. As a general rule to avoid potential statistical distortions, comparisons of the number of ships in the Navy over time should use, whenever possible, a single counting method.

In: U.S. Navy Shipbuilding
Editor: Gladys Clemens

ISBN: 978-1-63117-113-0

Chapter 2

NAVY SHIPBUILDING: OPPORTUNITIES EXIST TO IMPROVE PRACTICES AFFECTING QUALITY*

United States Government Accountability Office

WHY GAO DID THIS STUDY

The Navy expects to spend about $15 billion per year to provide its fleet with the most advanced ships to support national defense and military strategies. Problems with recently delivered ships have focused attention on quality issues.

House Report No. 112-110, accompanying the Department of Defense Appropriations Bill, 2012, mandated that GAO review the Navy's quality assurance processes for new ship construction. This report discusses, among other issues, (1) quality problems in constructing recently delivered ships and Navy actions to improve quality and (2) key practices employed by leading commercial ship buyers and shipbuilders to ensure quality and how these compared with Navy practices.

GAO analyzed Navy data on ship quality from 2006 to May 2013 and spoke with Navy officials and shipbuilders. GAO also reviewed deficiency data for commercial ships and spoke with buyers and builders.

WHAT GAO RECOMMENDS

To improve the construction quality of ships delivered to the Navy, GAO is recommending, among other things, that the Navy clarify policy on when deficiencies should be addressed, provide guidance on contract quality requirements, and assess applicability of certain commercial practices to Navy shipbuilding. DOD agreed with two recommendations and partially agreed with three, stating for example that current policy is adequate but that the

* This is an edited, reformatted and augmented version of the Highlights of GAO-14-122, a report to congressional committees, dated November 2013.

Navy would monitor deficiency trends. GAO believes that the recommendations remain valid as discussed in the report.

WHAT GAO FOUND

The Navy has experienced significant quality problems with several ship classes over the past several years. It has focused on reducing the number of serious deficiencies at the time of delivery, and GAO's analysis shows that the number of deficiencies—particularly "starred" deficiencies designated as the most serious for operational or safety reasons—has generally dropped. Nonetheless, the Navy continues to accept ships with large numbers of open deficiencies (see figure below as an example; although total deficiencies have declined for this ship class, the last ship still had about 1,000 deficiencies that the shipbuilder was responsible for correcting). Accepting ships with large numbers of uncorrected deficiencies is a standard practice and GAO found that there are varying interpretations of Navy policy with regard to when the defects should be resolved. In 2009, the Navy organization that oversees ship construction launched the Back to Basics initiative to improve Navy oversight of ship construction. However, a key output of the initiative promoting consistent and adequate quality requirements in Navy contracts has yet to be implemented.

Although the environment in which leading commercial ship buyers and builders operate differs in many ways from the Navy's, some commercial practices aimed at helping to ensure that ships are delivered with a minimum number of deficiencies may be informative for the Navy. Throughout the course of commercial shipbuilding projects, significant numbers of quality defects and instances of non-conforming work are identified. However, leading commercial ship buyers and shipbuilders make great efforts to ensure that these issues are resolved prior to delivery. Further, commercial ship buyers establish clear lines of accountability and hold their personnel responsible for ensuring the shipbuilder delivers a quality vessel. While commercial ship buyers focus on regularly witnessing in-process work through roaming patrols and impromptu inspections, Navy processes at the shipyards place less emphasis on in-process work. Moreover, leading commercial shipbuilders have strong quality management processes that track quality problems to the worker or supervisor level. Navy shipbuilding contractors have historically experienced difficulties in holding production workers and supervisors accountable for their work, but some of the shipyards reported they are making progress on increasing worker accountability.

ABBREVIATIONS

ABS	American Bureau of Shipping
DOD	Department of Defense
FPSO	Floating Production, Storage and Offloading vessel
INSURV	U.S. Navy Board of Inspection and Survey
LCS	Littoral Combat Ship
NAVSEA	Naval Sea Systems Command
NAVSEA 02	Naval Sea Systems Command Contracting Directorate

NAVSEA 04	Naval Sea Systems Command Logistics, Maintenance & Industrial Operations Directorate
NAVSEA 04Z	Naval Sea Systems Command Logistics, Maintenance & Industrial Operations Directorate, SUPSHIP Management Group
NAVSEA 05	Naval Sea Systems Command Engineering Directorate
NAVSEA 07	Naval Sea Systems Command Undersea Warfare Directorate
NAVSEA 08	Naval Sea Systems Command Nuclear Propulsion Directorate
NAVSEA 21	Naval Sea Systems Command Surface Warfare Directorate
PEO	U.S. Navy Program Executive Office
SUPSHIP	Naval Sea Systems Command Supervisor of Shipbuilding, Conversion and Repair
TSM	Technical Support Management data system

November 19, 2013

Congressional Committees

The Navy seeks to provide its fleet with the most advanced ships to support national defense and military strategies, expecting to spend about $15 billion per year building ships. Given the difficult operating environments and extended deployments for Navy ships, it is essential that they operate as expected. Yet several cases of poor quality in Navy shipbuilding programs have focused attention on quality issues for newly constructed ships. Many of the problems were attributed to issues with basic elements of shipbuilding, such as welding, installation of key systems (like propulsion and anchoring systems), and electrical work. The impact of poor quality can directly affect operational missions; for example, the first ship built in the USS *San Antonio* class (LPD 17) had to undergo emergency repairs during its first deployment that were primarily attributed to poor workmanship and a lack of quality control during the ship's construction. Recognizing that quality problems in shipbuilding needed to be addressed, the Navy established the "Back to Basics" initiative in 2009 to ensure the efficiency and quality of ship construction. The initiative focused on the Navy's Supervisor of Shipbuilding, Conversion and Repair (SUPSHIP)—the organization responsible for overseeing ship construction processes—and involved senior Naval Sea Systems Command leadership and the Navy Program Executive Offices responsible for managing the development and procurement of ships.

Also in 2009, we identified best practices from the commercial shipbuilding industry and made several recommendations to the Department of Defense to improve management of shipbuilding programs involving, among other things, knowledge needed at key decision points.[1] Building on that review, House Report No. 112-110, accompanying the Department of Defense Appropriations Bill, 2012 (H.R. 2219), mandated that we review the Navy's quality assurance processes for new ship construction. This report assesses (1) the extent to which newly constructed ships delivered to the Navy from 2006 through May 2013 had quality problems and the actions the Navy has taken to improve quality; (2) key practices employed by leading commercial ship buyers and shipbuilders to ensure satisfactory quality and the extent to which Navy shipbuilding programs employ these practices; and (3) the role of classification societies (e.g., the American Bureau of Shipping) in Navy and commercial shipbuilding.

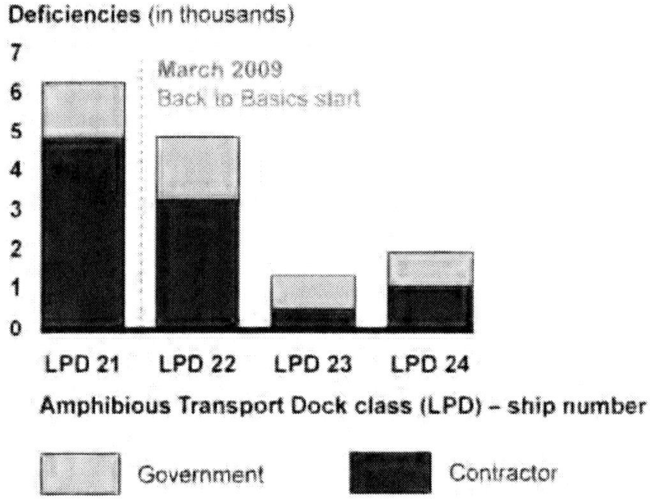

Source: GAO analysis of Navy data.

Deficiencies at Time of Delivery for LPD 17 Ship Class.

To identify the extent to which newly constructed Navy ships had quality problems and the actions the Navy has taken to improve quality, we reviewed Navy inspection reports, internal Navy reviews with regards to ship quality, ship delivery reports, shipbuilding contracts, and other documents discussing the quality of ships delivered to the Navy from 2006 through May 2013. We reviewed data on all ships delivered during this period but only compared trends in quality from those ship classes where multiple ships were delivered during this time period. To determine the number and type of deficiencies for each vessel, we obtained and analyzed data from the Navy's Board of Inspection and Survey's (INSURV) Material Inspection data warehouse and the Navy's Technical Support Management (TSM) system. TSM is the primary database SUPSHIP uses for tracking the status of new construction deficiencies. We reviewed these data for completeness, and when we identified obvious discrepancies we brought them to the attention of Navy officials and worked with them to understand, correct, or omit the discrepancies. We determined that the deficiency data we obtained were sufficiently reliable for the purposes of this report with two exceptions. These exceptions relate to data for T-AKE class ships. TSM data did not cover T-AKE 1 through T-AKE 6. In addition, data for T-AKE 12 had numerous data errors and is therefore not reported on. For other T-AKE ships, we reviewed deficiency documentation (trial cards) to resolve discrepancies between TSM and T-AKE program office data.

We visited eight U.S. private shipyards that build Navy ships and spoke with shipyard representatives. We also met with officials and analyzed data provided by several Navy organizations, including each of the SUPSHIP commands and detachments; INSURV; Program Executive Offices and shipbuilding programs; lifecycle and maintenance organizations; Fleet Forces Command; the Military Sealift Command; Navy Sea Systems Command Engineering, Contracting, and Logistics directorates, among others. We catalogued several hull, mechanical, and electrical quality problems with each ship class delivered since 2006. To create this list of illustrative examples, we asked Navy officials and shipbuilding contractor representatives to identify quality problems on these vessels. Further, this list focused only on quality issues that pertain to the construction of the hull, mechanical, and electrical systems; we did not include quality issues with weapons systems or other warfighting systems. We also reviewed the Navy's Back to Basics initiative and outcomes, as well as other recent efforts to improve the quality of shipbuilding.

To learn about practices used by leading commercial ship buyers and shipbuilders to ensure quality in new construction vessels, we spoke with leading buyers and shipbuilders in the cruise, oil and gas, and commercial shipping industries and reviewed our previous shipbuilding best practices work. Where possible, we collected documentation and/or witnessed quality assurance practices. For the purposes of this review, the leading commercial ship buyers we spoke with are companies that we identified as leaders in their industry in terms of being top operators of cruise ships, oil and gas vessels or containerships, and that agreed to participate in our review. We reviewed such indicators as annual sales, number of vessels owned or procured, and total market share. Leading commercial shipbuilders in this review were also identified as high quality shipbuilders by the ship buyers in our review or shipbuilding experts we met with. The firms participating in our review included Carnival Corporation, Chevron Corporation, Daewoo Shipbuilding and Marine Engineering, Ensco plc, ExxonMobil, Hyundai Heavy Industries, A.P. Moller-Maersk A/S, Meyer Werft, Noble Corporation, Norwegian Cruise Line, Royal Caribbean Cruises, Ltd., Seadrill Ltd., and STX Finland. We requested deficiency data from the commercial ship buyers for one or more new construction ships they acquired. With the exception of one floating production storage and offloading vessel, all of these ships were delivered to the buyers in 2012 or 2013. We assessed the reliability of these data by obtaining information on the systems that stored the data and interviewing ship buyer representatives knowledgeable about the data. We determined that the data were sufficiently reliable for the purposes of this report. We also identified common processes and tools used by these ship buyers and shipbuilders to ensure the expected level of quality. To determine the extent to which Navy quality assurance processes used commercial best practices, we reviewed data and information obtained from the Navy and its shipbuilding contractors as well as from the leading commercial ship buyers and shipbuilders. We also held meetings with SUPSHIP, program, and contracting officials about the Navy's quality practices and during our site visits to the U.S. shipyards. We discussed with shipyard representatives their quality assurance processes and the steps taken to ensure ships meet the Navy's quality expectations.

To better understand the role of classification societies in Navy and commercial shipbuilding, including the American Bureau of Shipping (ABS), we met with engineering and marine surveying representatives from ABS at the Navy contractor shipyards where they maintain a presence to obtain an overview of how they conduct their work. We also held discussions with officials from several Navy organizations, including SUPSHIP and Naval Sea Systems Command Engineering Directorate, as well as representatives from Navy shipbuilding contractors on the role of ship classification in Navy shipbuilding. We reviewed the classification rule set developed by the Navy and ABS for the Navy's surface combatants, as well as other classification rule sets pertaining to Navy and commercial vessels. We also spoke with representatives from other classification societies, including Det Norske Veritas and Lloyd's Register, to discuss their approach to classification of commercial and navy vessels, and met with commercial ship buyers and shipbuilders to discuss the classification process.

Appendix I contains additional detail on our scope and methodology.

We conducted this performance audit from March 2012 to November 2013 in accordance with generally accepted government auditing standards. Those standards require that we plan and perform the audit to obtain sufficient, appropriate evidence to provide a reasonable basis for our findings and conclusions based on our audit objectives. We believe that the evidence obtained provides a reasonable basis for our findings and conclusions based on our audit objectives.

BACKGROUND

In general, a quality product is one that performs as expected and can be depended on to perform when needed. A quality product is also one that is free of deficiencies. For the purposes of this review, we define deficiencies as items that require corrective action to bring the material condition or performance of a product into compliance with required standards.[2]

Performance, cost, schedule, and quality goals are interrelated. Improving quality within an organization to achieve the desired level of performance and reliability requires time and money in the short term. But in the longer term, such efforts can reduce costs and production times, as process- related costs—such as labor and materials needed to correct defects— are reduced or eliminated. Such production efficiencies have been demonstrated by companies that have successfully implemented quality philosophies, including the Six Sigma methodology.[3] Six Sigma and other quality philosophies establish a framework by which the process of production can be made more consistent by determining and eliminating the root causes of process-related problems, resulting in improved quality. Figure 1 demonstrates the interrelationship between schedule, cost, performance and quality.

Shipbuilding is a major undertaking within the Navy, and several Navy organizations are involved, either directly or indirectly, with the acquisition, construction, and fielding of new ships. To some extent, all of the organizations described below have a role in helping to ensure the Navy acquires ships that meet quality expectations.

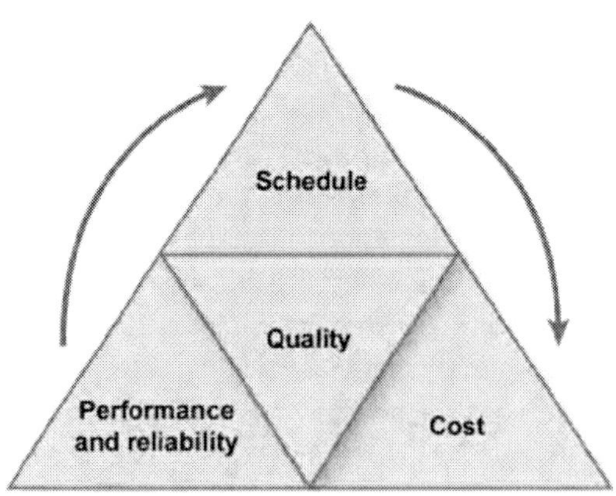

Source: GAO analysis of American Society of Quality data.

Figure 1. Quality's Integral Role in Achieving Schedule, Cost, and Performance and Reliability Goals.

- **SUPSHIP** is the Navy's primary on-site representative at the private shipyards that build Navy ships, and is tasked with overseeing the shipbuilder's production processes. Services provided by SUPSHIP include contract administration, project management, engineering surveillance, quality assurance, logistics, and financial administration of assigned shipbuilding contracts. Typical work activities performed by SUPSHIP's quality assurance department include the following:

    - Quality assurance planning activities, such as oversight plans for monitoring the shipbuilder's quality program and periodic surveillance plans identifying the areas where quality assurance personnel will be allocated.
    - Reviews of the shipbuilding contractor's quality management system and work procedures.
    - Inspection and testing of the shipbuilder's completed work, including physical inspections, verifications, and equipment testing, as well as witnessing or monitoring the ship construction process.
    - Audits and inspections of the shipbuilder's work procedures to verify that personnel are complying with the procedures.
    - Evaluating the results of (1) SUPSHIP's quality inspections to identify quality trends and (2) quality and test data the shipbuilder is contractually required to provide the Navy.
- **Program Executive Offices (PEO),** and the program managers that report to them, are responsible for all aspects of life-cycle management of their assigned shipbuilding programs, including program initiation, ship design, construction, testing, delivery, fleet introduction, and maintenance activities.
- **The Navy's Board of Inspection and Survey (INSURV)** is an independent organization that inspects newly constructed and in- service Navy ships to determine their material condition and reports these assessments to Congress and Navy leadership.
- **The Naval Sea Systems Command Contracting Directorate (NAVSEA 02)** awards contracts worth about $24 billion annually for new construction ships and submarines, ship repair, major weapon systems, and services.
- **The Naval Sea Systems Command Engineering Directorate (NAVSEA 05)** provides the engineering and scientific expertise and technical authority for the Navy's ships, submarines, and associated warfare systems.
- **The Naval Sea Systems Command Nuclear Propulsion Directorate (NAVSEA 08)** is responsible for all aspects of the Navy's nuclear propulsion ships, including research, design, construction, and operations.
- **The Military Sealift Command** operates and maintains approximately 119 non-combatant Navy ships that deliver supplies to US forces and conduct specialized missions, such as supporting humanitarian aid efforts. Unlike other Navy ships, the ships are generally manned by civilian mariners. During construction of these vessels, the Command assigns a construction oversight representative at the shipyard to monitor compliance with requirements.

- **The Naval Sea Systems Command Surface Warfare Directorate (NAVSEA 21)** and **Undersea Warfare Directorate (NAVSEA 07)** provide life-cycle support for the Navy's surface combatants and submarines from when they are introduced into the fleet until they are no longer in service. In addition, the directorates provide the shipbuilding and submarine program offices with information on quality problems and lessons learned from ship deployments that may be used to improve the construction of subsequent vessels.
- **The Defense Contract Management Agency** administers contracts when delegated that authority by the contracting office. SUPSHIP commands generally delegate quality oversight of shipbuilding parts and equipment suppliers to the Defense Contract Management Agency.

Figure 2 depicts how those Navy organizations fit within the overall organizational structure of the Navy.

Stages of Shipbuilding

Shipbuilding is a complex, multistage industrial activity that includes a number of key events that are common regardless of the type of ship constructed or whether the buyer is the government or a commercial firm. These events are sequenced among four primary phases: pre- contracting, contract award, design and planning, and construction, with each successive phase building upon the work completed in earlier stages.[4]

In the pre-contracting stage, the ship buyer determines the ship's requirements. Early-stage design work occurs that culminates in a set of specifications that documents the buyer's requirements. In the contract award phase, the specifications are incorporated into the shipbuilding contract that the buyer enters into with the selected shipbuilder(s). After entering into the shipbuilding contract, the design and planning phase begins. The ship's detailed engineering design is completed—often in the form of a three-dimensional computer aided design model—and consists of developing all aspects of the ship's structures and the routing of major distributive systems, such as electrical work or piping. Any necessary modeling and simulation analyses, such as testing the structural integrity of the design over the service life of the ship, are also completed. In addition, during this stage the shipbuilder plans for beginning construction and generates two-dimensional drawings that, once approved by the ship buyer, will be used by production workers to build the ship.

Ship construction involves the following key events:

- **Block fabrication, assembly, outfitting, and erection:** Metal plates are welded together into elements called blocks—the basic building units for a ship. Blocks are generally outfitted with pipes, brackets for machinery or cabling, ladders, and any other equipment that may be available for installation at this early stage of construction. The blocks are then welded together to form grand blocks and erected with other grand blocks in a drydock or building area. Outfitting work, painting, and pre-commissioning activities take place prior to sea trials.

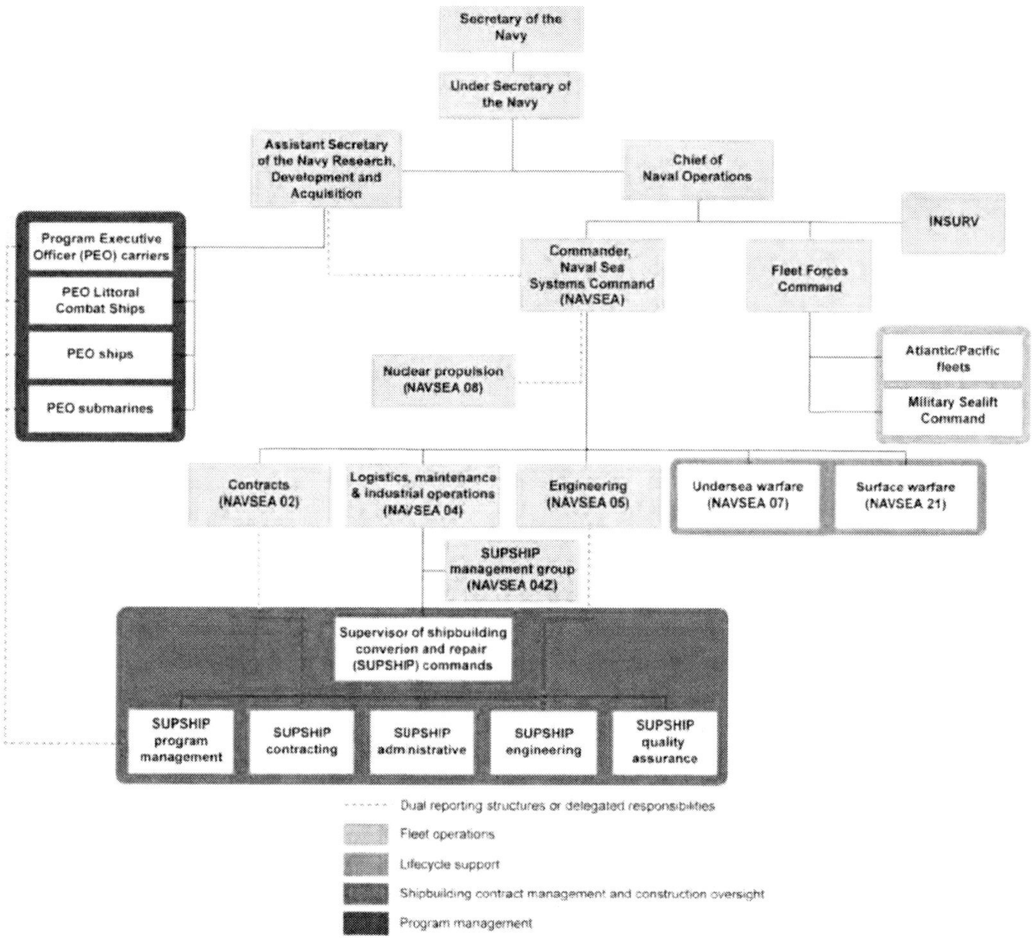

Source: GAO analysis of Navy data.

Figure 2. Simplified Organizational Chart of Navy Organizations Involved in the Construction of New Navy Ships.

- **Sea trials:** Once the shipbuilder is satisfied that the ship is completed, the ship embarks on a series of dockside and at-sea tests to evaluate overall quality and performance against the contractually required technical specifications and buyer's performance requirements. Navy shipbuilding programs generally conduct two sets of sea trials—builder's trials and acceptance trials.[5] SUPSHIP inspectors are generally responsible for observing and identifying deficiencies during the builder's sea trials, while acceptance trials are observed by inspectors from INSURV, the Navy's inspection board.
- **Delivery / acceptance:** Following the successful completion of sea trials and when the buyer is satisfied that the ship meets requirements, the shipyard delivers the ship to the buyer. In Navy shipbuilding, the official transfer of custody occurs when the Navy signs a Material Inspection and Receiving Report (Form DD 250).

Post-delivery activities that are specific to Navy shipbuilding, which can generally take up to a year to complete, include the following:

- **Final outfitting, post-delivery tests:** Following delivery and until the ship sails away from the shipbuilder's yard—usually anywhere from 10 to 90 days after delivery—the crew boards the ship and begins training, and the ship's mission systems are installed. Additional training and operational tests of mission systems occur at the ship's home port.[6]
- **Final contract trials:** INSURV inspectors conduct a second round of sea trials to assess whether the ship and all mission equipment are operating as intended.
- **Post Shakedown Availability:** A period of planned maintenance follows final contract trials. During this time, class-wide upgrades and correction of new or previously identified deficiencies that are the government's responsibility also occur.
- **Obligation and Work Limiting Date:** The official date on which full responsibility for funding the ship's operation and maintenance is transferred from the acquisition command to the operational fleet.

Construction-related quality deficiencies on Navy ships can be identified at all points throughout the shipbuilding process, from initial block fabrication to sea trials up through delivery. During acceptance trials, INSURV inspectors label the most serious issues as "starred" deficiencies. These issues can significantly degrade a ship's ability to perform an assigned primary or secondary operational capability or the crew's ability to safely operate and maintain ship systems. Because of their importance, starred deficiencies must be corrected by the builder or waived by the Chief of Naval Operations prior to ship delivery.

In addition to starred deficiencies, INSURV inspectors categorize other types of deficiencies—each of which represent a piece of the vessel that is not in compliance with Navy standards and/or contract specifications at the time of inspection. While not deemed as serious as starred deficiencies, these items can nonetheless affect the quality of life and safety of the sailors on board the vessel or the operability of the ship. INSURV categorizes these issues into three parts based on the professional judgment of its inspectors:

- **Part I deficiencies** are very significant in that they are likely to cause the ship to be unseaworthy or to substantially reduce the ship's ability to carry out its assigned mission. "Part I Safety" is a sub-category that INSURV uses to indicate that an issue is severe enough that the ship is unsafe to operate until corrected. All starred deficiencies are Part I deficiencies but not all Part I deficiencies are starred. An example of a Part I starred deficiency would be an anchor that when deployed during testing was not fully retrievable.
- **Part II deficiencies** involve less significant material degradation but should be corrected to restore the ship to required specifications. Part II deficiencies can also have a safety designation. Examples are wide ranging and can include items such as missing signage, or areas of the ship having missing or damaged paint and coatings.
- **Part III deficiencies** are generally categorized as things that prevent the ship from meeting Navy standards but are cost prohibitive to fix. An example is a lifeboat compartment that is too small to fit the size of a lifeboat necessary to meet Navy requirements.

While INSURV inspectors identify deficiencies during acceptance trials, throughout the ship construction process SUPSHIP quality inspectors, during the normal course of their work, may observe work being performed that is not in accordance with the technical specifications, quality requirements included in the contract between the Navy and the shipbuilder, or the shipbuilder's work procedures. In such instances, SUPSHIP inspectors will issue a request for the builder to correct the deficient work. Depending on the severity and extent of the problems, SUPSHIP's quality officials may send a request for corrective action to the builder to determine the cause of the problems, correct the affected work, and improve its work processes so that the problems will not reoccur. Once the shipbuilder takes the appropriate actions, SUPSHIP closes out the corrective action request. At the time of the builder's sea trials, any unresolved SUPSHIP requests for corrective action are generally grouped together with deficiencies identified by SUPSHIP inspectors during the builder's sea trials process.[7]

Navy Shipbuilding Environment

As opposed to the commercial buyers included in our review, which typically operate in a robust, competitive environment, the Navy has a limited industrial base to build its ships. For example, throughout the world there are at least 12 shipbuilding and offshore marine companies that ship buyers in the oil and gas sector can choose from to build their vessels. In contrast, two U.S. Navy shipbuilding contractors—General Dynamics and Huntington Ingalls Industries—own all but three of the larger shipyards, and each yard is specialized to build specific types of ships. For example, aircraft carriers can only be built at one location. In addition to building Navy ships, two of the shipyards, General Dynamics NASSCO and V.T. Halter Marine, also build vessels for commercial shipping firms. Figure 3 depicts the major U.S. shipbuilders and the types of ships they build.

The Navy uses several different types of contracts for shipbuilding programs. The Navy will often use cost-reimbursement contracts for the first ships of a new class, as in many instances the Navy and its shipbuilding contractors do not have a full understanding of the costs and effort needed to build and deliver the ships. Such contracts provide for payment of allowable incurred costs, to the extent prescribed in the contract. The Navy also typically includes incentive or award fees on these contracts. Our prior work has shown that this type of contract places more cost risk on the government in the event that the shipbuilder is unable to deliver the ship on time and within budget.[8] For shipbuilding programs where the Navy has greater certainty about costs and risks, the Navy typically employs fixed-price-incentive contracts. Fixed-price- incentive (firm target) contracts include a target cost, target profit, ceiling price (maximum price) and a formula used to determined the shipbuilder's profit that are negotiated at the outset. Final costs that are above the target cost but below the ceiling price are shared between the Navy and shipbuilder through the profit adjustment formula. If the final costs exceed the ceiling price, the shipbuilder is generally responsible for most additional costs. In some instances the Navy will use a firm-fixed-price contract, in which the final price of the ship is agreed to at the outset. This contract type generally places upon the contractor maximum risk and full responsibility for all costs.

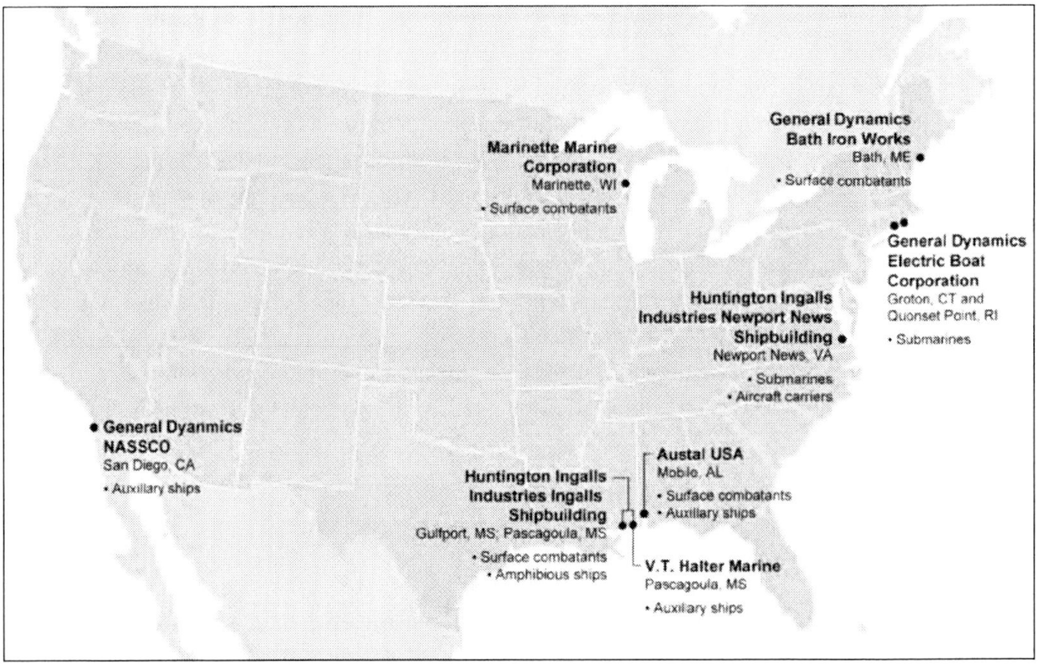

Source: GAO analysis of Navy and shipbuilding contractor data.
Note: Auxiliary ships include transport, cargo, and ammunition ships.

Figure 3. Locations of Major Navy Contractor Shipyards and Associated Product Lines.

Classification Societies

The maritime industry has certain requirements to ensure ships meet a minimum level of safety and quality. Under the International Maritime Organization conventions, uniform requirements have been established to, among other things, ensure safety of life while at sea and environmental protection.[9] These requirements stipulate that ships are designed, constructed, and maintained in accordance with the rules of a recognized classification society or with applicable national standards that provide an equivalent level of safety. Classification societies, such as the American Bureau of Shipping (ABS), Det Norske Veritas, and Lloyd's Register, develop rules defining a minimum level of technical standards that are applied to ships. Once a ship is "classed" with a certificate indicating that it meets a minimum level of safety and quality, the ship is subject to periodic inspection to verify that it continues to meet the applicable rules of the issuing classification society. In many instances, shipbuilders also use classification societies to audit their quality management systems as an independent third-party, which often is needed to maintain certain certifications such as the International Organization for Standardization's ISO 9001 (quality), 14001 (environmental), and 18001 (occupational health and safety) series of management systems standards.[10]

Many of these international shipping requirements regarding classification do not apply to Navy ships.[11] However, in some instances the Navy voluntarily complies with certain maritime and commercial classification requirements, such as for ships operated by the Military Sealift Command.[12] Beginning in 2003, the Navy entered into an agreement with ABS to assist with redefining the Navy's standards for the design and construction of its non-nuclear surface combatant ships—surface ships that are designed to engage in attacks against land, air, and sea targets—to be more consistent with the classification process used on commercial and Military Sealift Command-operated ships. The result of this effort was development of the Naval Vessel Rules, which establish a minimum set of requirements for the basic construction of the Navy's surface combatant ships.[13] In addition, the Navy required shipbuilding contractors to contract with ABS to pilot the application and implementation of the Naval Vessel Rules on the USS *Zumwalt* (DDG 1000) and Littoral Combat Ship (LCS) shipbuilding programs. Under this arrangement, ABS involvement in the LCS and DDG 1000 programs is similar to that of a commercial shipbuilding project in which, throughout design and construction of the ship, the classification society acts as an independent third-party assessor to ensure the ship is in compliance with the applicable classification rule sets. Upon completion of all requirements, ABS issues a classification certificate.

The Navy Has Reduced Deficiencies at Delivery but Still Accepts Some Ships with Numerous Construction Deficiencies

Recognizing that it has experienced significant quality problems with several ship classes, the Navy has focused on reducing the number of serious deficiencies, particularly "starred" deficiencies, which require a waiver from the Chief of Naval Operations to defer correction until after delivery. The number of uncorrected deficiencies at delivery, including starred deficiencies, has generally dropped for ships delivered over the past few years due, at least in part, to the Back to Basics quality improvement initiative. Even so, the Navy still accepts some ships with large numbers of open deficiencies. Navy policy states that ships are to be delivered based on acceptance trials and satisfactory correction or resolution of deficiencies. Instead, correction of deficiencies is often deferred until after the Navy accepts delivery of the ship, which can interfere with post-delivery activities. Further, one product of the Back to Basics initiative, a quality management provision for inclusion in Navy shipbuilding contracts, has not been implemented on any Navy shipbuilding contract.

The Navy Has Reduced the Number of Starred Deficiencies at Delivery

The Navy has reduced the total number of uncorrected starred deficiencies at delivery on both established and newer ship classes for which multiple ships have been delivered. It is up to INSURV inspectors to categorize deficiencies identified by severity during acceptance trials. Because there can be ambiguity regarding who is responsible for correcting the deficiencies, the program office, SUPSHIP, and the shipbuilder collectively determine whether the government or the shipbuilding contractor is responsible. Deficiencies that the government is responsible for correcting can include, among other things, problems requiring a change to the ship design or ship specification, or equipment that the government is responsible for providing. For example, during the acceptance trial of LCS 2, the rescue boat could not be properly deployed and INSURV inspectors categorized the problem as a starred deficiency. It was subsequently determined that, as designed, the rescue boat system did not have the ability to meet requirements and that the Navy would assume responsibility for the corrective work. Figures 4 through 7 show the open number of starred deficiencies at delivery for the LPD 17, DDG 51, T-AKE, and LCS ship classes during the 2006 through 2012 time frame.[14] These are the classes of ships that had multiple vessels delivered in this time frame. The figures also show when the Back to Basics initiative began.

In addition to starred deficiencies, total deficiencies have declined for several ship classes. Notably, the last ship of the T-AKE class, T-AKE 14, was delivered in 2012 with no deficiencies that the shipbuilder was responsible for correcting. Uncorrected deficiencies were also kept to a minimum on the recently delivered Joint High Speed Vessel and Mobile Landing Platform—both based on commercial designs and operated by the Military Sealift Command. The first Joint High Speed Vessel, an intratheater troop and cargo transport ship, was delivered in December 2012 with only 54 uncorrected deficiencies, of which 6 were categorized as Part I deficiencies. The first Mobile Landing Platform, which will provide at-sea cargo and equipment transfers, was delivered in May 2013 with only three uncorrected shipbuilder-responsible deficiencies, according to the Navy. These had not been resolved prior to delivery because the shipbuilder was waiting for parts to correct the items.

Figure 8 shows the number of open deficiencies at the time of delivery for the T-AKE ship class. The figure also shows when the Back to Basics initiative began. In addition, appendix III provides additional information on the numbers and types of deficiencies at different points in time for T- AKE class ships delivered to the Navy from 2009 through 2012.

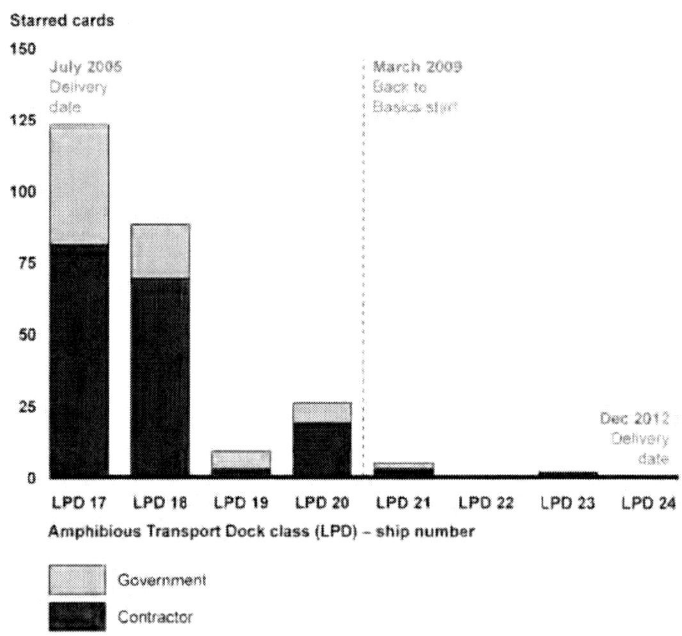

Source: GAO analysis of Navy data.

Note: LPD 17 was delivered to the Navy in 2005, but is included in our analysis as this was the lead ship for the San Antonio class.

Figure 4. Open Starred Deficiencies at Delivery for LPD 17 Ship Class.

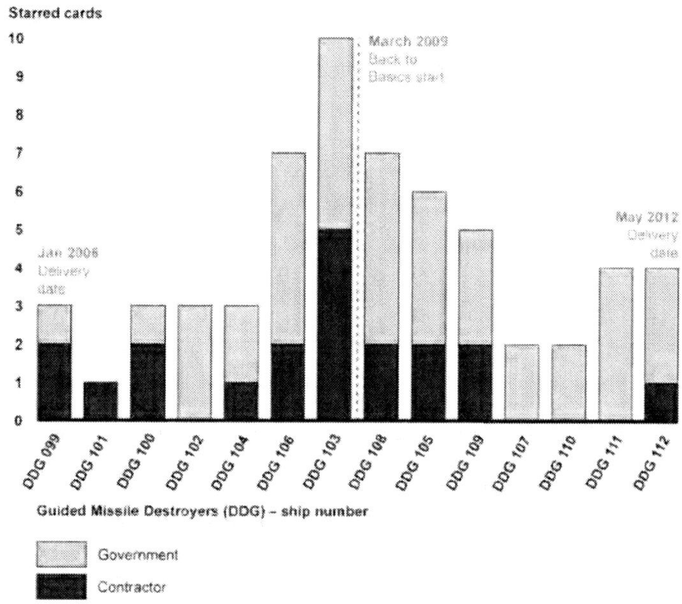

Source: GAO analysis of Navy data.

Note: DDG 51 class ships are built at two Navy contractor shipyards in Bath, Maine and Pascagoula, Mississippi. Ships are presented in chronological order based on the date of delivery to the Navy.

Figure 5. Open Starred Deficiencies at Delivery for DDG 51 Ship Class.

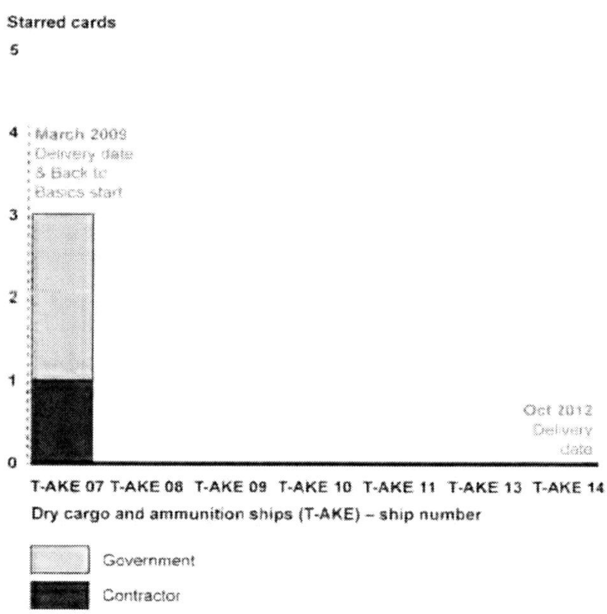

Source: GAO analysis of Navy data.

Note: The database used for this analysis did not contain information on T-AKE 1 through T-AKE 6, delivered between 2006 and 2008. Data obtained for T-AKE 12 indicated that almost all of the deficiencies were opened and closed on the same day following the ship's delivery and, as such, data for that ship were deemed not sufficiently reliable for the purpose of this analysis.

Figure 6. Open Starred Deficiencies at Delivery for T-AKE Ship Class.

It should be noted that a confluence of several factors led to improved quality for the T-AKE and Mobile Landing Platform ship classes. The quality of ships generally improves as ship classes mature. Production efficiencies are gained from process improvements and incorporation of lessons learned into the build strategy. During early production of the T-AKE ship class, the shipbuilder initiated shipyard-wide production improvements and other efficiencies resulting, in part, from the shipbuilder's partnership with a leading international commercial shipbuilder. With the improvement in quality, the T-AKE program moved to a contract type which transfers more of the cost risk resulting from any quality problems to the shipbuilder. Specifically, for T-AKE 10 through T- AKE 14 the Navy transitioned from a fixed-priced-incentive contract to a firm-fixed-price contract. The shipbuilder for the Mobile Landing Platform attributed the delivery of a nearly defect-free ship to its focus on ensuring that all design engineering efforts and production planning activities were fully completed prior to the start of the lead ship's construction. Shipbuilder representatives stated that this was a key factor in keeping rework rates for quality problems low.

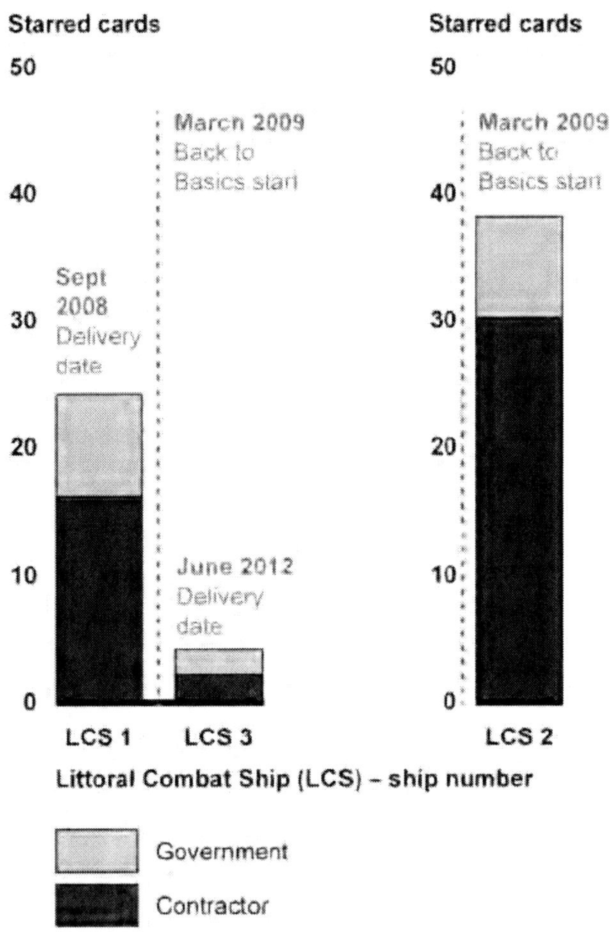

Source: GAO analysis of Navy data.

Note: Odd numbered LCS class ships (*Freedom* variant) are built at a shipyard in Marinette, Wisconsin, while even numbered ships (*Independence* variant) are built at a shipyard in Mobile, Alabama.

Figure 7. Open Starred Deficiencies at Delivery for LCS Ship Class.

Numbers of Other Deficiencies at Delivery Remains Substantial

Even with the drop in the number of starred deficiencies, the Navy has continued to accept delivery of some ships with large numbers of uncorrected deficiencies. For example, LPD 22, which was delivered without any starred deficiencies and was cited by Navy officials as a turning point for the LPD 17 ship class, had over 3,300 deficiencies that the contractor was responsible for correcting at delivery. Subsequent LPD ships, LPD 23 and 24, both delivered in 2012, had fewer but still sizeable numbers of uncorrected deficiencies. On these ships, examples of uncorrected Part II deficiencies that were the shipbuilder's responsibility when delivered included pipe hangers that were insufficiently spaced to support the weight intended or that did not meet the Navy's requirements; fire suppression sprinklers not providing adequate coverage due to obstructions and interference from pipes and ductwork; and inability to fully extend a crane boom due to overhead obstructions on LPD 24. While Part II deficiencies are considered to be not as severe as Part I deficiencies, these deficiencies nonetheless can require a fair amount of shipbuilder effort to remedy.

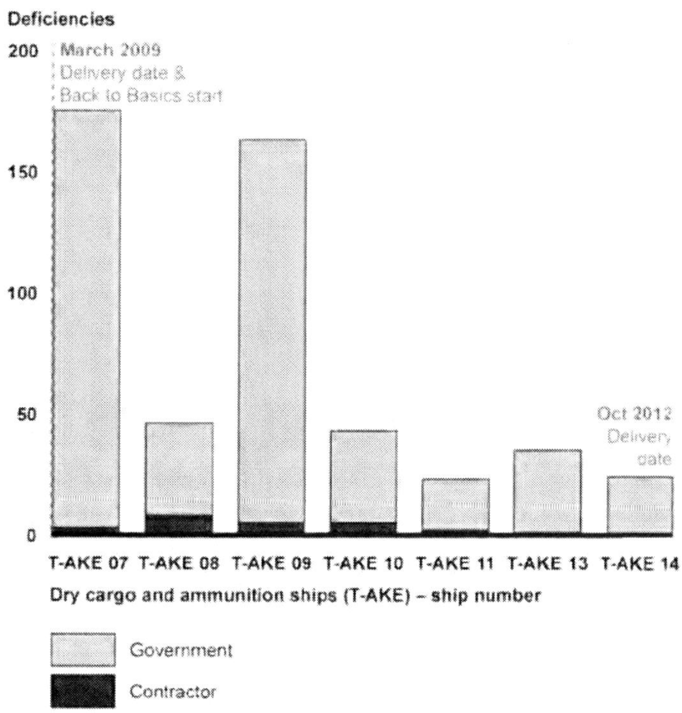

Source: GAO analysis of Navy data.

Notes: The database used for this analysis did not contain information on T-AKE 1 through T-AKE 6, delivered between 2006 and 2008. Data obtained for T-AKE 12 indicated that almost all of the deficiencies were opened and closed on the same day following the ship's delivery and as such, data for that ship were deemed not sufficiently reliable for the purpose of this analysis. In some instances, similar types of deficiencies were consolidated into a single deficiency prior to ship delivery. The data above contains deficiencies where it was subsequently determined no further corrective action would be taken.

Figure 8. Non-Starred Deficiencies at Time of Delivery for T-AKE Ship Class.

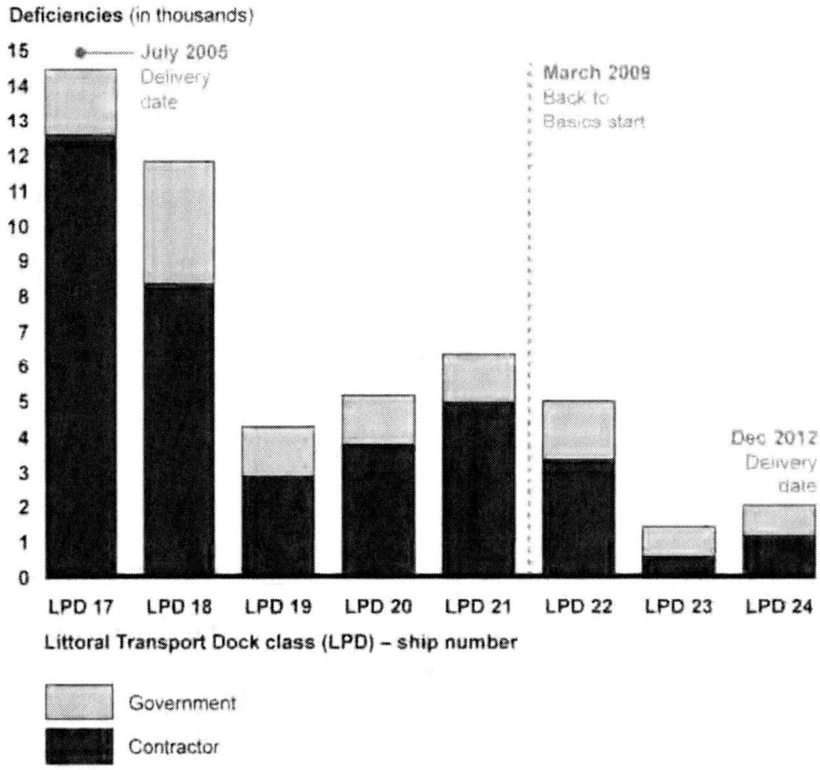

Source: GAO analysis of Navy data.

Notes: LPD 17 was delivered to the Navy in 2005, but is included in our analysis as this was the lead ship for the San Antonio class.

Deficiencies closed up to 7 days after date of delivery are treated as being closed at delivery to account for potential lags in data entry.

Figure 9. Non-Starred Deficiencies at Time of Delivery for LPD 17 Ship Class.

Figure 9 shows the total number of open non-starred deficiencies (Part I, Part II, and Part III) at the time of delivery for the LPD 17 ship class. The figure also shows when the Back to Basics initiative began. Appendix IV provides additional information on the numbers and types of deficiencies at different points in time for LPD class ships delivered to the Navy since 2009, when the Back to Basics initiative began.

As another example, on LCS 3, uncorrected Part II deficiencies that were the shipbuilder's responsibility at the time of delivery included use of incorrect weld-filler material on pipe joints associated with the ship's water jets; instances where cabling was insufficiently supported, was bent, or had insufficient banding; and valves that were inaccessible due to obstructions.

Figure 10 summarizes the number of open non-starred deficiencies (Part I, Part II, and Part III) at the time of delivery for the LCS ship class. The figure also shows when the Back to Basics initiative began. In addition, appendix V provides additional information on the numbers and types of deficiencies at different points in time for the LCS ship class.

In the above examples for the LPD and LCS ship classes, many of the deficiencies were first identified by SUPSHIP inspectors either prior to or during the builders' sea trials but were not corrected before delivery.

In addition, correcting very significant deficiencies (Part I) can require various levels of effort. For example, INSURV designated a broken window in the helicopter control station as a Part I deficiency on DDG 112. This may be relatively easy to fix, but is labeled a Part I deficiency because, according to deficiency documentation, until this issue is corrected the aviation facilities on the vessel cannot be fully certified, limiting operations. However, other issues could be more complex to correct. On DDG 108, for example, INSURV inspectors identified locations within the ship where there was insufficient corrosion protection, which if left uncorrected could result in accelerated corrosion as it did on at least one previous hull. Additional corrosion protection features were necessary to correct the issue, resulting in changes to the ship specification for follow-on hulls. Figure 11 summarizes the number of open non-starred deficiencies (Part I, Part II, and Part III) at the time of delivery for the DDG 51 ship class. The figure also shows when the Back to Basics initiative began. Appendix VI provides additional information on the numbers and types of deficiencies at different points in time for those DDG 51 class ships delivered to the Navy since 2009, when the Back to Basics initiative began.

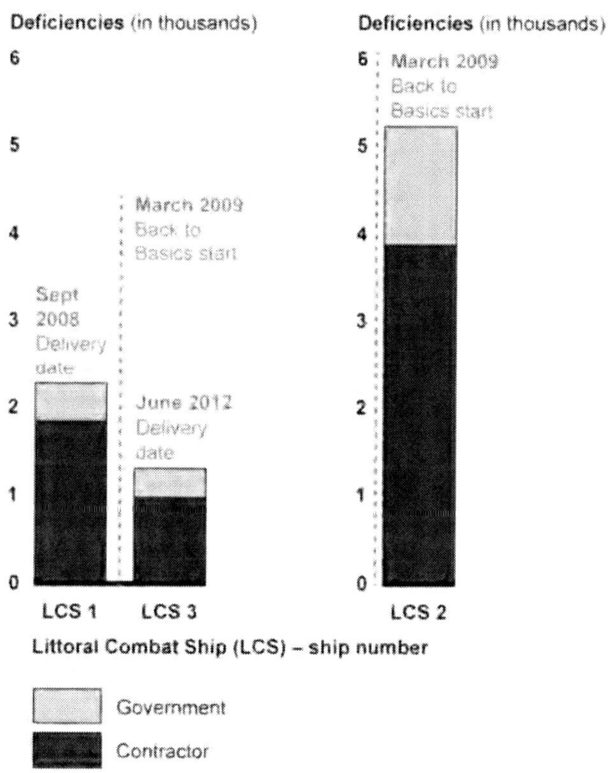

Source: GAO analysis of Navy data.

Notes: Odd numbered LCS ships (*Freedom* variant) are built at a Navy contractor shipyard in Marinette, Wisconsin, while even numbered ships (*Independence* variant) are built at a Navy contractor shipyard in Mobile, Alabama.

Deficiencies closed up to 7 days after date of delivery are treated as being closed at delivery to account for potential lags in data entry.

Figure 10. Non-Starred Deficiencies at Time of Delivery for LCS Ship Class.

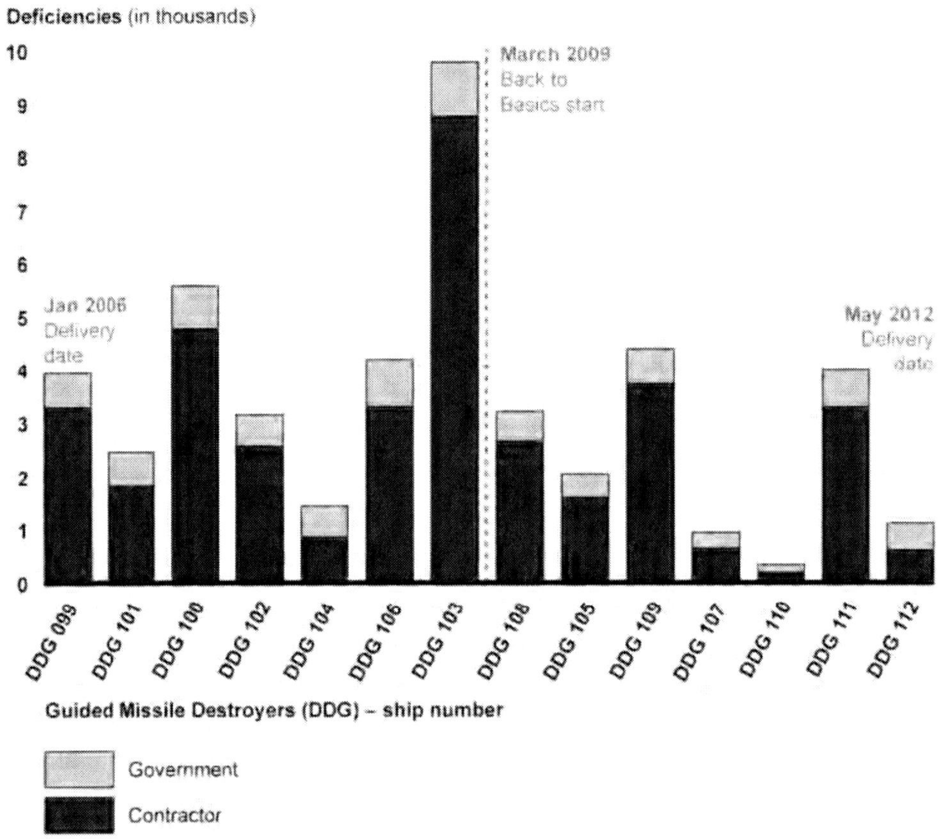

Source: GAO analysis of Navy data.

Notes: DDG 51 class ships are built at two Navy contractor shipyards in Bath, Maine and Pascagoula, Mississippi. Ships are presented in chronological order based on the date delivered to the Navy. Deficiencies closed up to 7 days after date of delivery are treated as being closed at delivery to account for potential lags in data entry.

Figure 11. Non-Starred Deficiencies at Time of Delivery for DDG 51 Ship Class.

While our analysis indicates that deficiencies have generally been decreasing where there are multiple ships in a class, we found instances where ships built individually or infrequently experienced difficulties in meeting the Navy's requirements prior to being delivered to the Navy. These include the *Wasp* class amphibious assault ship (LHD 8), the *Nimitz* class aircraft carrier (CVN 77), and a missile instrumentation ship (T-AGM 25).

- The Navy accepted delivery of LHD 8 in April 2009 with over 12,000 uncorrected deficiencies which were determined to be the shipbuilder's responsibility. Of these, almost 70 percent were for items SUPSHIP inspectors had identified before or during the builder's sea trials, including numerous deficiencies related to the ship's cabling and wiring. According to representatives of the shipbuilder, following delivery the shipbuilder undertook an intensive effort lasting several months to assess the condition of the ship's cabling, as it was unclear which cables and wires were properly routed. As a result of the costs incurred by the shipbuilder for this effort, the builder completely revised and modernized the approach used to install cables and wires on subsequent ships built at the shipyard.

- According to the shipbuilder's list of incomplete work items and deficiencies, the Navy took delivery of CVN 77 in May 2009, with about 7 percent of the ship's compartments unfinished; over 8,000 uncorrected deficiencies that had been identified prior to or during sea trials; and approximately 3,900 generally minor deficiencies (such as missing cable tags and electrical outlets) that were identified by the crew during its inspection. In addition, INSURV inspectors noted during CNV 77's acceptance trial that because the ship was not fully primed and painted, longer-term cost and maintenance implications needed to be addressed, as corrosion developed in the unfinished spaces and machinery.
- During the May 2011 acceptance trial for T-AGM 25, INSURV inspectors recommended that the Navy not take delivery of the ship until an additional acceptance trial was conducted because almost none of the major systems on the vessel were in satisfactory condition, and several were incomplete or inoperable. The inspectors also noted that most of the deficiencies they cited were previously identified during builder's trials but had not been properly corrected and retested. In the months following, the shipbuilder conducted an additional builder's trial to demonstrate some key components along with INSURV's second acceptance trial. At the time of delivery in January 2012, there were 67 unresolved deficiencies from sea trials which the shipbuilding contractor was responsible for correcting, and about 600 additional items identified by the Navy requiring correction by the builder, including improperly secured and supported cables, damaged paint, missing safety equipment, and loose tiles.

In these cases, the Navy gave priority to programmatic goals and determined it was in the best interest of the Navy to proceed with key milestones despite a large volume of uncorrected deficiencies or incomplete work. According to Navy officials, the Navy may make such determinations based on its strategic needs or additional costs that may be incurred by not proceeding with key milestones. For example, the LHD 8 program office took delivery of the ship as it was over a year and half behind schedule, primarily due to lingering effects from Hurricane Katrina, a shipyard-wide strike, and delays completing the ship's machinery control system. With CVN 77, SUPSHIP and program office officials stated it was necessary to take delivery of the ship in its unfinished condition so that it could periodically participate in training missions while the remaining work was being completed. Lastly, program officials decided to proceed with acceptance trials for T-AGM 25 to meet scheduling constraints associated with installation of the ship's mission package at another location, which could only be accomplished after taking delivery of the ship. In addition, SUPSHIP officials noted that in some instances, such as with T-AGM 25, the decision was made to proceed with sea trials in an effort to reveal additional problems with the ship that may not have been known at the time.

Deficiencies May Persist into Operations

Deficiencies identified by SUPSHIP and INSURV, if not sufficiently corrected, may link to problems later when the ship is in operation. In some instances, issues identified by SUPSHIP's quality inspectors during construction were closed prior to delivery—indicating the builder had satisfactorily addressed the problems—but the root causes of the problems were not fully resolved. For example, during the construction of DDG 107 issues with welding were identified in the normal course of SUPSHP's quality inspections. Specifically, SUPSHIP inspectors identified weld defects on the ship's mast and issued a request for corrective action to the shipbuilder. The action was closed, indicating that the builder had resolved the identified problems. However, during a routine training exercise in February 2011, a sub-section of the ship's mast failed when the welds could no longer support the structure. According to Navy investigators, it is likely that the section of the mast would have fallen onto the ship or into the water, but cables prevented it from falling. Although the section of the mast that failed was not at the same location where defective welds were identified by SUPSHIP, a Navy investigation into the issue identified several causes of the failure, including inadequate quality assurance oversight by the shipbuilder and poor workmanship during the construction process.

In addition, deficiencies can also combine to form larger issues. Quality problems that stem from deficiencies are often complex and can manifest long after the error occurred that led to the problem. Navy documents addressing quality concerns state that hull, mechanical, and electrical issues are often the result of one or more of the following causes:

- Construction and workmanship deficiencies, such as weld defects or inadequate painting;
- Design errors, such as refueling at sea systems that cannot accommodate necessary fuel offloading rates; and
- Supplier and subcontractor issues, such as counterfeit parts and sub- systems that do not work properly.

For example, a gas turbine engine on LCS 1 was ruined and had to be replaced because it was flooded with sea water. The Navy determined that this occurred because (1) the welds for the doors and water separators were not flush, creating gaps that allowed water to pass through the doors (a workmanship issue); (2) there were gaps in the sealing surfaces of the water separator that also allowed water to penetrate (a design issue); and (3) a key filter became clogged because it was not replaced in a timely manner (a maintenance issue).

Figure 12 catalogues select quality problems for a number of ships related to hull, mechanical, and electrical workmanship that persisted following delivery of the vessels. To generate this list, we asked Navy officials from many different departments, including program offices, maintenance units, and ship operators, to identify problems facing the various ship classes built in the last eight years.

Accepting a Vessel with Deficiencies Is Common, and Varying Interpretations of Policy Dictate Practice

Policy issued by the Office of the Chief of Naval Operations (*Navy Instruction 4700.8J, Trials, Acceptance, Commissioning, Fitting Out, Shakedown, and Post Shakedown Availability of U.S. Naval Ships Undergoing Construction or Conversion*) states the following:

- For new construction ships, all contractual and governmental responsibilities should be resolved prior to delivery, except for crew certification, outfitting, or special Navy range requirements which cannot be met until after delivery.
- Delivery of the ship is based on acceptance trials and satisfactory correction or resolution of deficiencies, and acceptance trials shall be conducted when all work, including the correction of significant known deficiencies, has been completed.
- In many cases it may be prudent to defer work until the post-delivery period before the vessel is transferred to the fleet, for example for financial or workload reasons.

In addition, the Navy's fixed-price shipbuilding contracts that we reviewed included the Navy's Delivery of Completed Vessel clause, which requires the shipbuilder, before the ship is delivered to the Navy, to satisfactorily correct all contractor responsible deficiencies as necessary to avoid an adverse effect on the operational capability of the vessel.[15]

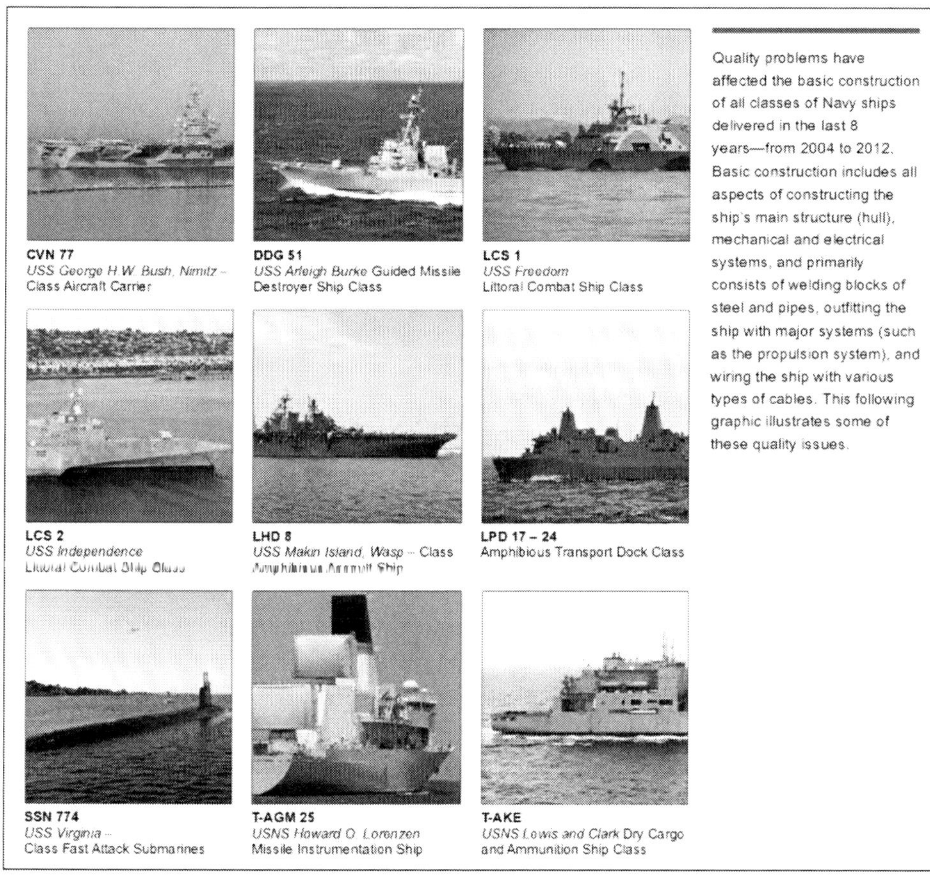

To see more detailed information about each ship see Appendix VII.
Source: GAO analysis of Navy data (data and images).

Figure 12. Shipbuilding Issues.

Different units in the Navy disagree over when ships should be free from deficiencies. The Office of the Chief of Naval Operations, the authors of the policy, state that the policy is to accept delivery of a vessel from the shipbuilder that is free of deficiencies, especially all contractor- responsible deficiencies. However, program office and SUPSHIP officials, as well as officials from other Navy organizations, stated that it is standard practice for the Navy to accept delivery of a ship with numerous unresolved deficiencies and/or incomplete work. Navy program officials point to the part of the policy that states that the vessel should be fully mission capable by the Obligation and Work Limiting Date—the time full financial responsibility for the ship is transferred to the operational fleet—and this is their quality goal. Thus, the Navy has routinely deferred correcting deficiencies with the intention of correcting these items during the post-delivery period, as shown in appendixes III through VI.[16] In fact, for some, such as those in the DDG 51 class, the bulk of deficiencies are addressed during the first four months following delivery to the Navy during the final outfitting period. This is the time when the crew boards the ship and begins testing and training prior to the ship being deployed on a mission. Several NAVSEA officials stated that addressing deficiencies during the post-delivery period can interfere with crew training, final outfitting, and testing of the vessel while also affecting the quality of the work being performed.

Further, while some Navy officials told us that transferring problems to the fleet is rare, the transfer of major problems to the fleet in varying degrees of severity occurred on LPD 17-21, LHD 8, and DDG 103, requiring the use of operations and maintenance funds to correct the defects. Navy operations officials told us that following the problems with LPD 17 and LPD 18 that were transferred to the fleet, the fleet has become more involved in ensuring that it receives a quality vessel. They noted that the fleet now has the ability to provide input at key points during the acquisition process on whether identified problems on a vessel have been adequately addressed.

Navy Has Implemented Some, but Not All, Quality Improvement Measures

In 2007, recognizing the widespread quality problems in new construction ships, the Commander of NAVSEA began to examine the quantity of SUPSHIP's quality assurance inspectors in a few key locations to determine what resources were needed to improve quality. Following this workforce assessment, between fiscal years 2009 and 2012, the Navy hired new staff and reallocated funding for engineering, acquisition management, and quality assurance personnel at those SUPSHIP commands with the greatest risks to quality and in order to keep up with the growth of shipbuilding programs.

In 2009, the Back to Basics effort was initiated. This effort identified several quality assurance related goals and developed a means to help SUPSHIP better communicate to the program offices the role of quality assurance and the purpose of quality oversight. In addition, NAVSEA standardized many of the operating procedures across all SUPSHIP locations. Another focus of Back to Basics was improving SUPSHIP's oversight of critical hull, mechanical, and electrical shipbuilding processes such as welding, painting, and cabling by standardizing and improving its use of metrics to assess shipbuilder performance in these areas. For example, on a quarterly basis, the SUPSHIP commands report to NAVSEA leadership on the extent to which the shipbuilders are meeting SUPSHIP's quality goals as compared with previous periods. These quarterly reports provide NAVSEA leadership with insight into the reasons behind any increases in quality problems.

The Back to Basics initiative prompted at least two SUPSHIP locations to use a process called "pulse audits" where SUPSHIP and shipbuilder quality inspectors conduct an inspection together to ensure that their inspections are consistent. Also, some SUPSHIP locations have regular meetings with quality inspectors from the shipbuilder to compare quality inspection results and metrics, and discuss any discrepancies between the results of their inspections. As a result, SUPSHIP and shipbuilder officials reported they now have a greater understanding of each other's quality assurance processes.

In addition, in June 2010, the Back to Basics team developed a quality performance standard that set forth common quality requirements to be included in shipbuilding contracts, which was subsequently published as a NAVSEA technical publication. The Quality Performance Standard for Construction of Naval Vessels provides standard contract language that, among other things, would require shipbuilders to

- develop and submit a quality assurance plan for the government's approval;
- conduct a review to identify special controls, processes, equipment, and skills required for assuring product quality;
- ensure that drawings are reviewed for adequacy and completeness before use;
- have a quality system that ensures that work is inspected and/or tested at points necessary to ensure conformance with contract requirements;
- respond to corrective action requests within 21 days (7 days for safety issues) and indicate when corrective action will be completed;
- maintain and use cost data on prevention and detection of defects and failure costs (such as scrap, rework, and repairs);
- provide the government electronic access to quality, accuracy control, and manufacturing process data; and
- use results-oriented indicators to demonstrate quality program effectiveness.

Although the standard was developed in 2010, according to officials who wrote it, as of September 2013 none of the Navy's shipbuilding programs had incorporated this quality performance standard into their shipbuilding contracts. Further, in a May 2013 meeting, senior NAVSEA leadership told us they were unaware that a quality performance standard had been developed, noting that quality clauses are already included in the Navy's shipbuilding contracts. According to officials from SUPSHIP and the SUPSHIP Management Group, the quality standard was developed because of inconsistencies in quality requirements included in shipbuilding contracts that decreased the Navy's ability to effectively conduct oversight. For example, we reviewed two fixed-price-incentive shipbuilding contracts awarded in fiscal year 2011 that contain very different quality requirements. On one contract for a relatively mature ship class, the quality clause includes production process control, data sharing, and the use of results-oriented metrics to demonstrate quality program effectiveness. On another contract for a newer ship class, the quality clause simply requires that the shipbuilder develop, implement, and maintain a quality assurance plan covering certain aspects of shipbuilding.

Further, SUPSHIP officials cited one recent contract that did not include provisions for SUPSHIP to review the shipbuilder's work procedures. As a result, the builder was reluctant to provide this information as it was not contractually required, limiting SUPSHIP's ability to discover potential problems. SUPSHIP officials noted that, under current contracts, the range of data provided may not be sufficient for them to do their work. For example, officials noted that one fixed-price-incentive contract stated that "processes or indicators (internal design changes, production rework, etc.) to be monitored and reported shall be agreed to by the Program Office and shall be identified in the management plan(s);" without going into any additional detail about the specific types of data or the frequency with which data was to be provided.

Views regarding the success of Back to Basics are mixed. NAVSEA leadership views it as a success, noting that maintaining the improvements made over the last few years will be critical as budgets shrink and the shipbuilding portfolio shifts into more steady-state production. They point to the LPD 17 and T-AKE programs as evidence of the benefits resulting from the quality assurance staffing increases. However, others within the Navy believe that quality improvements on some shipbuilding programs, such as the T-AKE program, were largely independent of the initiative.

COMMERCIAL FIRMS RESOLVE QUALITY DEFICIENCIES BEFORE DELIVERY, WITH SOME PRACTICES POTENTIALLY INFORMATIVE FOR NAVY IN SPITE OF DIFFERENT ENVIRONMENT

The environment in which leading commercial ship buyers and builders operate differs in substantial ways from the Navy's. Key differences include the level of competition, different incentives at play, and the timing of ship deliveries to the end user. Nevertheless, some commercial practices supporting delivery of ships with a minimum number of deficiencies may be useful for the Navy. These practices include

Source: Royal Caribbean Cruises Ltd.

Celebrity Solstice.

- a focus on resolving deficiencies before ship delivery,
- contracting approaches that place the cost risk associated with addressing quality problems on the shipbuilder and incentivize prompt resolution of problems, and
- an oversight process with clear lines of accountability and an emphasis on observing in-process work.

Environment and Key Incentives Differ between Commercial and Navy Shipbuilding

The commercial shipyards and ship buyers we visited build and buy ships that are highly complex and densely outfitted. These ships include floating production storage and offloading (FPSO) vessels, which collect, process, and store oil from undersea oil fields; large cruise ships, some of which are comparable to the size of an aircraft carrier and can accommodate over 5,000 passengers; liquefied natural gas carriers; and offshore drilling ships, which in some instances can sit unanchored and drill for oil in water depths of over 10,000 feet while maintaining a fixed position. Construction can take up to three years at a cost ranging from about $600 million for a drill ship to well over $1 billion for a cruise ship or FPSO. In many cases, these ships incorporate technological advances that are vital to improving business operations. For example, for enhancing operational and commercial efficiencies, the new generation liquefied natural gas carriers developed a range of techniques from reliquefaction of the boil off gas to its utilization as fuel for engines for propulsion. Key characteristics of selected commercial ships and offshore structures included in our review are presented in appendix VIII.

Commercial shipbuilders, particularly those in the oil and gas industry, operate in a robust, competitive environment, as opposed to the U.S. Navy's limited shipbuilding industrial base, where sole source contracts may be awarded in order to sustain workloads and the solvency of the companies involved. This environment provides commercial buyers with additional leverage to expect quality at delivery, as they can go elsewhere if they are not satisfied with the end product. One oil and gas company reported that there are as many as a dozen shipbuilders from which they can solicit proposals at any one time. As a result, commercial shipbuilders put a premium on reputation. They do not want to risk their credibility with buyers if the ship is not delivered by the contracted delivery date with the buyer's expected level of quality. These factors create additional pressure on the shipbuilder to ensure that outstanding deficiencies are resolved in a timely manner. In contrast, the Navy has fewer choices of shipbuilders and has an interest in sustaining these shipbuilders despite shortfalls in performance.

In addition, commercial buyers and builders operate in an environment where both parties seek to maximize profits. A delay in delivery has significant profit impacts to both the buyer and shipbuilder. For example, it is common for a drill ship to be leased to an oil and gas company early during the ship's construction process, and both the ship buyer and the company leasing the vessel rely on the ship to be at its drill site by the contracted date to begin operations. Any delay in meeting this schedule can not only cause the ship buyer to lose revenue—which for some drill ships can amount to over $600,000 per day—but also may require the buyer to pay predetermined sums to the company leasing the drill ship for lost work days. Also, newly constructed cruise ships are expected to start generating revenue weeks if not days after being delivered. For example, Royal Caribbean officials told us that they had a full cruise 3 days after taking delivery of the *Celebrity Solstice* in October 2008. The monetary risks faced by the ship buyer for late delivery or acquiring a ship that does not meet key performance requirements—such as fuel consumption—are also passed along to the shipbuilder in the form of liquidated damages that can potentially cost the shipbuilder tens of millions of dollars.[17]

Leading Commercial Ship Buyers Focus on Taking Delivery of Ships Meeting Quality Expectations

Throughout the construction process, the ship buyer's oversight team, the shipbuilders' quality personnel, and classification society personnel routinely identify deficiencies such as design errors, supplier and vendor quality issues, and problems with workmanship. Commercial ship buyers we met with expect that identified problems will be corrected prior to delivery of the ship. We found this to be the case for all types of ships included in our assessment of commercial practices, whether a lead ship (i.e., the first ship built according to a new design) or a ship built from a proven design. Unresolved issues that affect the safety, seaworthiness, or operability of the ship would be reason for the ship buyer to not accept delivery until such items are corrected. For the generally minor deficiencies that have not been corrected by the ship's delivery date, the buyer and shipbuilder may enter into a formal agreement outlining specific actions that the builder will take to ensure correction of the outstanding issues in the prescribed time.

There is a fair amount of subjectivity as to what constitutes a major or minor deficiency that largely depends on the ship's intended use. While each ship-buying company may take a slightly different view of what constitutes a major or minor deficiency, minor deficiencies are generally those items that do not have an effect on the mission, operability, or safety of the vessel and are not indicative of problems with the builder's production processes. Major deficiencies can be viewed as problems with the shipbuilder's production processes that limit the builder's ability to meet ship specifications, or those deficiencies that could have an adverse effect on the mission, operability, or safety of the vessel if not resolved. We found that commercial ship buyer definitions for major and minor deficiencies do not necessarily align with the definitions used by the Navy. For example, in the cruise industry, where aesthetics and the customer experience are critical factors, a major deficiency may be an issue such as higher than anticipated engine vibration or elevated noise levels. Typical minor deficiencies include cosmetic defects such as scratches on windows, paint, and furniture. For oil and gas ships that are expected to endure harsh environments, common major outstanding deficiencies at the time of delivery could involve equipment that has not yet been installed or equipment that requires additional calibration. Typical minor deficiencies might include missing signage or technical manuals.

A drill ship for Noble Corporation, the *Noble Don Taylor*, illustrates the process for correcting deficiencies leading up to delivery of commercial ships. In this case, the shipbuilder reduced the number of corrective items, totaling over 15,000 throughout construction, to around 3,800 by the time sea trials were scheduled to occur. At the time of ship delivery, the number of unresolved corrective actions was down to just over 59 items, which the shipbuilder agreed to correct no later than three months following the date of delivery. One day following ship delivery, this number dropped to 37 items. These items included minor deficiencies such as incorrect labels and missing manufacturer documentation for equipment, and in some instances, more significant corrective actions including replacement of equipment which measures the weight and torque of the drill that did not have the correct calibration.

Additional examples depicting the extent to which quality problems identified by the ship buyer are resolved prior to key delivery milestones are provided in table 1.

The commercial ships reviewed were delivered with considerably fewer defects than is common with Navy ships, even with the recent improvements realized by the Navy. With the exception of the Celebrity cruise ship where there were a large number of minor (often cosmetic) uncorrected deficiencies at the time of delivery, the number of total deficiencies ranged from 4 to 59 for the commercial ships reviewed. For the Navy, recent T-AKE class ships, the Joint High Speed Vessel, and the Mobile Landing Platform were comparable to the numbers of outstanding deficiencies found on commercial ships.

Commercial Ship Buyers Put More Risk on Shipbuilders through Choice of Contract Type and Payment Structure

The Navy and commercial ship buyers agree that responsibility for quality must be put on the shipbuilders, as they are in the best position to ensure quality. However, the commercial buyers we spoke with structure their contracts to ensure that the shipbuilder absorbs the cost risks associated with quality problems. For example, they require delivery of a ship at an expected quality level for a firm-fixed-price and delay a majority of payment until the expectation is met. The Navy shares more of the cost risk associated with delivery of a quality product with its shipbuilders through cost-reimbursement[18] and fixed-price-incentive[19] contracts. The Navy also makes regular progress payments throughout construction so that the bulk of payment has already been made by delivery.

Source: Noble Corporation.

Nobal Corporation's *Noble Don Taylor* dynamically positioned drill ship.

Table 1. Unresolved Deficiencies for Select Commercial Ships at Key Delivery Milestones

Ship name (type)	Delivery date	Prior to sea trials		After sea trials		At ship delivery	
		Open minor deficiencies	Open major deficiencies	Open minor deficiencies	Open major deficiencies	Open minor deficiencies	Open major deficiencies
Noble Don Taylor (drill ship)	April 2013	3,803	7	1,839	6	37	22
Seadrill Ltd. West Auriga (drill ship)	April 2013	387	78	455	105	50	0
Chevron Big Foot (tension leg platform production facility – hull structure only)a	December 2012	436	89	16	26	a	a
Ensco DS 6 (drill ship)	January 2012	~100	0	<100	0	4	0
Celebrity Cruises, Inc. (subsidiary of Royal Caribbean Cruises, LTD.) Reflection (cruise ship)	October 2012	9,110	30	8,100	25	680	20
Star Deep Water Petroleum, LTD (a Chevron-affiliated company) *Agbami* – Floating Production, Storage and Offloading vessel (FPSO)	June 2008	2,051	150	164	10	15	0

Source: GAO analysis of industry-provided data.

[a] This project used a unit rate contract where the shipbuilder is paid a specific rate for performance of the work that is proportional to the volume of work needed to complete the project. The contractor that is integrating the production facility onto the hull of the platform, rather than the shipyard, will address the open deficiencies. Therefore the ship buyer could not provide us with the data.

Commercial Ship Buyers Use Firm Fixed-Price Contracts to Put Cost Risk of Poor Quality on the Shipbuilder

Firm fixed-price contracts used in commercial shipbuilding—even for lead ships—put more of the cost risks associated with quality on the shipbuilder than cost-reimbursement and fixed-price-incentive contracts which are commonly used by the Navy. Under a firm-fixed-price contract, the shipbuilder takes on the full cost of any quality problems that result in rework. As many quality problems require rework to existing blocks or compartments and can thus erode the shipbuilder's profit that is included in the firm-fixed-price, there is a greater incentive to minimize production deficiencies throughout construction.

Similar to commercial ship buyers, the Navy has made some limited use of firm-fixed-price contracts to purchase ships included in our review, and in those cases the contractor has taken on more cost risk associated with any quality problems. Of the 11 shipbuilding programs we reviewed, three (DDG 51, T-AKE, and T-AGM 25) used firm-fixed-price contracts.[20] For example, starting with the 10th ship in the T-AKE ship class, the Navy successfully moved to a firm fixed-price contract for the remaining 5 ships. Whereas some of the earlier T-AKE ships exceeded the contracted target price, the ships constructed under firm-fixed-price contracts continued to maintain good quality. T-AGM 25 was purchased using a firm-fixed-price contracting arrangement from the outset, but in this case the project experienced quality problems and delays. Although the shipbuilding project did not perform as expected, the Navy's exposure to cost overruns resulting from quality problems was mitigated.

Under a cost-reimbursement contract, the cost risks associated with poor quality remain with the Navy because the government pays for all allowable costs of construction, including any rework, although lower costs may be incentivized by the use of award or incentive fees. The Navy has used cost-reimbursable contracts on lead ships due to concerns that the level of uncertainty and risk common on Navy programs make fixed-price contracts too costly. For example, one NAVSEA directorate stated that some Navy shipbuilding programs would likely be deemed unaffordable under a fixed-price contract due to the shipbuilders' need to include in the price the risk of uncertainty associated with new construction methods, new technologies, and new designs. However, the Navy generally moves more mature ship programs to fixed-price-incentive contracts, and Navy officials have stated that the cost sharing provisions in this type of contract incentivize quality. Under a fixed-price-incentive type contract, the contract is awarded for a target cost. The Navy and the shipbuilder share both cost savings and cost overruns below or above the target cost (referred to as the "share line"), pursuant to a formula, until the ceiling price is reached. Navy officials have stated that they see this type of contract as an incentive to quality because the shipbuilders will receive more profit if they construct the ship efficiently and deliver it below the target cost. Appendix IX further illustrates how cost risks pertaining to quality are allocated under the different contract types.

Commercial Shipbuilding Payment Terms Put More Pressure on Shipbuilder to Deliver Complete Ship at Expected Quality Level

In the commercial world, ship buyers use payment terms as leverage to ensure that the shipbuilder delivers a ship to the expected level of quality. Payments are generally made at milestones negotiated with the shipbuilder, such as contract signing, steel cutting, and keel laying. The bulk of the payment, sometimes 60 to 80 percent, is made only on delivery of a ship that meets expected quality and performance levels. The buyers we met with alter payment terms based on the quality of the shipbuilder and also use the payment terms to incentivize the builder to fix any outstanding deficiencies prior to making the last payment. For example, one construction manager stated that his company normally makes equal payments to shipbuilders at five different milestones, so that 20 percent of the ship cost remains to be paid at delivery, but the company would likely increase the percentage of the overall payment outstanding at delivery with a lesser quality shipbuilder. Another ship buyer said that his company normally pays 60 to 70 percent of contract price at delivery, but would not make final payment if the ship had major outstanding quality issues. Two other ship buyers told us they make milestone payments during construction, but generally retain 5 to 10 percent of the payment as a means to ensure that the builder addresses deficiencies before delivery. Those buyers also said they can retain these funds at delivery to ensure prompt resolution of any unresolved deficiencies. These practices create a strong financial incentive for the builder to quickly complete work and clear any outstanding quality defects. One project manager indicated that on two recent shipbuilding projects, it was not necessary to withhold any of the payment after taking delivery of the vessels because of the small number of deficiencies.

Of the 11 fixed-price contracts we reviewed (with the exception of TAGM- 25), the Navy makes periodic payments to its shipbuilders according to the progress made in construction.[21]

For example, one contract allows the shipbuilder to submit payment invoices every two weeks that are based on the progress made in construction as long as the billed amount is over $5,000. While commercial ship buyers may pay the bulk of payment at delivery, this system also requires shipbuilders to finance construction and related finance costs, which are ultimately passed on to the ship buyer in the contract cost.

According to Navy contracting officials, they consider potential impacts on pricing in developing payment provisions. In the fixed-priced shipbuilding contracts we reviewed, we found that the Navy generally retains a percentage from each progress payment—anywhere up to 10 percent— that decreases as the ship is being built. Navy contracting officials told us that they will only reduce the retention to the lower percentage once outstanding problems have been cleared. The following example from a recent shipbuilding contract illustrates this process:

- Up to 25 percent of the ship's physical completion—5 percent is retained from each progress payment;
- 25 to 50 percent of the ship's physical completion—3 percent is retained from each progress payment;
- 50 to 75 percent of the ship's physical completion—1.5 percent is retained from each progress payment; and
- Once 75 percent of the ship's physical completion has been reached—1 percent is retained from each remaining progress payment.

At delivery, the Navy retains a minimum amount of the shipbuilder's payment as a performance reserve that ranges from 0.75 to 1.5 percent of the ship's contract value, and will withhold additional funds from the shipbuilder's last invoice or the amount retained during construction if there are uncorrected deficiencies or incomplete work. However, the Navy may be constrained in terms of how it can incentivize contractors since the bulk of payments are made during the course of construction rather than at delivery. As it is common for Navy ships to have many more deficiencies and incomplete work items at delivery than the commercial ships in our review, the amount retained may not sufficiently motivate the shipbuilder to correct all deficiencies. Further, there is not specific NAVSEA guidance that addresses the extent to which retentions should be used as a means to incentivize the shipbuilder to promptly resolve outstanding deficiencies and incomplete work items at ship delivery. However, standards for internal control in the federal government identify the need for documenting policies and procedures to ensure appropriate measures are taken to address risk.[22] Program officials stated that for uncorrected deficiencies that are large in scope they will typically develop an estimated cost to complete each item. For the remaining deficiencies, the officials stated they develop an average cost that is based on the estimated cost to complete a selected sample of the deficiencies and apply that cost universally across the remaining deficiencies. In the case of LHD 8, which was delivered to the Navy in April 2009 with over 12,000 outstanding shipbuilder-responsible deficiencies, the Navy retained about 2.5 percent of the final estimated cost. Navy officials told us that some of these outstanding items are still being completed during maintenance periods.

More recently, Navy and SUPSHIP officials stated that in at least three instances they have temporarily increased the amount of the retained payments in an effort to prompt the shipbuilder to correct persistent deficiencies. SUPSHIP officials told us this approach was successful for problems they encountered with the LHA 6, DDG 51, and LPD 17 ship classes. In the case of LPD 17 class ships, the Navy withheld over $15 million in progress payments on four ships that were under construction until the shipbuilder resolved persistent problems with pipe cleanliness.

Commercial and Navy Shipbuilding Utilize Differing Approaches to Foster Accountability and Ensure Quality

Commercial Ship Buyer Oversight Structure Creates Clear Lines of Accountability While Navy Oversight Is More Diffused

Another mechanism commercial ship buyers use to ensure quality assurance is having dedicated, trained inspection teams on site to monitor and oversee all aspects of construction. All of the commercial ship buyers we met with create clear lines of responsibility for functional areas to ensure accountability within their on-site teams, which are primarily responsible for quality during construction. Within the on-site teams of the commercial buyers we met with, responsibility for a functional area such as paint or hull and structure is consolidated under one functional area lead who reports to the buyer's overall project or construction manager on-site. The on-site project manager has overall responsibility within the buyer's company to ensure that the shipbuilder delivers the ship on time and at the expected level of quality. Quality inspectors are integrated within each functional area team. Commercial ship buyers told us that their on-site quality inspectors are expected to independently identify critical inspection areas during the course of their day-to-day inspections. We found that commercial quality inspectors use design drawings and ship specifications during their work to ensure that the items are built in accordance with the detailed design. Furthermore, these inspectors are also responsible for tracking and closing all corrective actions they identify during the course of construction and ensuring that the corrected work or work processes resolved the identified problems. Overall, commercial ship buyers place a high-level of responsibility on their inspectors to be able to identify important quality issues in their functional area and ensure that the company will take delivery of a ship that meets quality standards.

Within the Navy, SUPSHIP quality assurance departments have limited authority over the shipbuilder compared to commercial ship buyers' inspectors. SUPSHIP quality assurance teams identify and report defects found during their inspections, and can close out the deficiencies once the defects are rectified. However, deficiencies that are technical in nature, or with which the builder disagrees, are adjudicated by SUPSHIP engineering and the ship's program office and not the quality assurance department. For example, SUPSHIP's engineering department reviews and adjudicates technical issues related to design and system engineering that quality assurance teams identify during inspections. If the engineering group finds that a defect is technically acceptable, a waiver may be provided indicating that the shipbuilder does not need to correct the defect. The ship's program office is primarily responsible for successful delivery of the ship, including responsibility for cost, schedule and performance requirements. Program office officials also review certain defects identified by the SUPSHIP quality assurance teams and make the determination as to whether or not they will be corrected by the shipbuilder, provided the defect is technically acceptable.

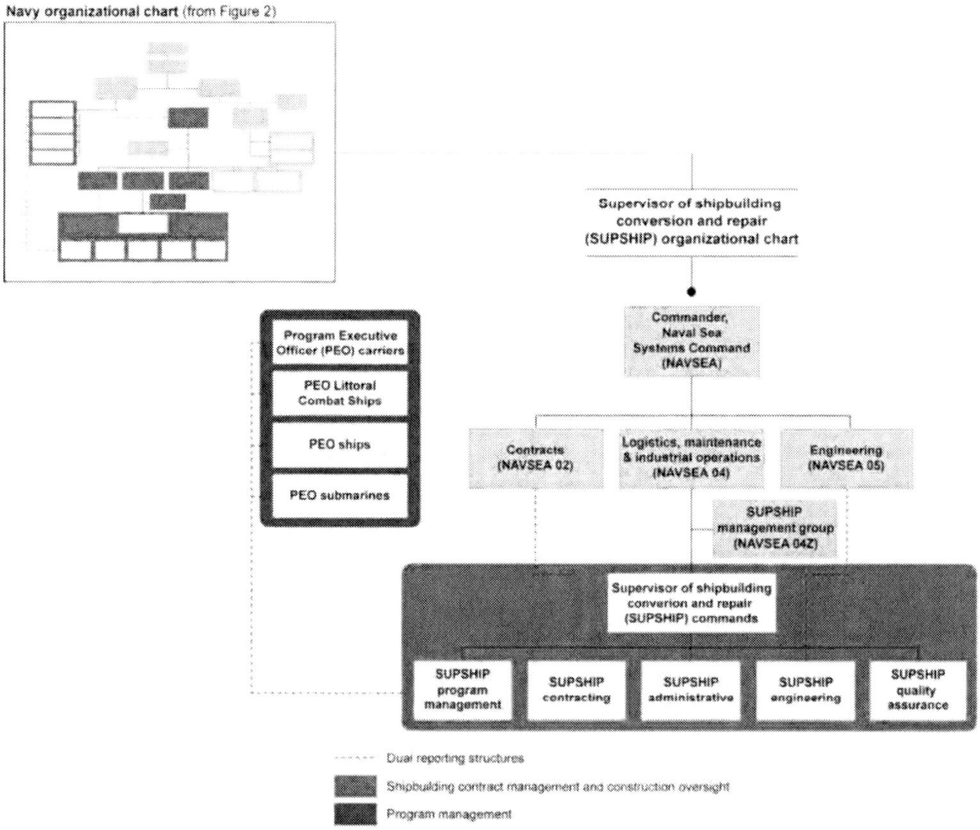

Source: GAO analysis of Navy data.

Figure 13. Simplified Organizational Chart of SUPSHIP Indicating Those Departments Having Delegated Responsibilities.

As shown in figure 13 below, there are major departments within SUPSHIP, such as contracting, engineering, and program management that have delegated responsibilities from the respective NAVSEA level directorates or the PEOs and are able to elevate technical risks or concerns about the structure of the shipbuilding contract to those that have decision-making authority. SUPSHIP quality officials receive policy guidance from the NAVSEA Logistics, Maintenance and Industrial Operations Directorate (NAVSEA 04), but there is not a quality assurance team at the NAVSEA level to which they can raise quality related issues.

Officials within NAVSEA 04 told us the Navy is formulating plans to reorganize the structure of the directorate, and plans include establishing a centralized quality team. Although the roles, responsibilities, and authorities of the quality team are not yet defined, the officials expect that the team would provide support to SUPSHIP quality assurance departments in helping to ensure that matters related to quality are given sufficient attention at the NAVSEA level. Increasing the emphasis on quality may help contribute to the goal of delivering ships that are defect- free, or nearly defect-free, as called for in Navy policy and demonstrated in commercial shipbuilding.

Commercial Ship Buyers Align Quality Inspections with Shipbuilder Plans While Navy Has Multiple Inspection Plans, Fewer In-Process Inspections

Prior to construction, the commercial ship buyers and shipbuilders we met with negotiate to one common understanding of quality inspections through an agreed-upon quality inspection plan.[23] The plan identifies (1) all formal inspection points during the construction process; (2) who attends and approves the inspections; and (3) criteria on how the inspections will be carried out. One ship buyer noted that establishing the quality inspection plan in conjunction with the shipbuilder and ensuring that all parties agree to, and are aware of, the key drivers that affect quality is of the utmost importance in ensuring the ship is built and delivered at the expected level of quality. According to company officials, the shipbuilder is contractually required to notify the buyer of any formal inspection points to provide the buyer an opportunity to inspect a product or process at a designated point of production.

Buyer representatives, including on-site teams and headquarter-level officials, review the builder's quality inspection plan and may ask for additional inspections and hold points based on ship design, criticality of the system, and previous experiences of the buyer. For example, one ship buyer reported experiencing significant failure rates during pressure testing of piping systems on a drill rig, and decided to incorporate additional inspection points at the joints where pipes are welded together on future drill ship projects. In another instance, a ship buyer representative told us his company solicits input on the quality inspection plan from company personnel that operate similar types of vessels, which he noted provides valuable information on the types of quality issues observed from an operator's perspective. The quality inspection plan is a key tool in ensuring quality as it enables consistency in inspections by focusing the ship buyer and builder on the same inspection items using the same criteria. Figure 14 further describes actions taken by one commercial firm, Chevron's Project Resources Company, to improve quality.

Inspectors for commercial ship buyers enforce quality requirements, in part, by attending all formal inspections, according to almost all of the commercial ship buyers in our review (seven of nine companies). For example, one ship buyer we met with reported having up to 80 inspections in a day for the team of about 14 inspectors. These inspectors use design drawings and review the ship specification to ensure that the compartment or block is built in accordance with requirements.

Chevron Corporation's Big Foot tension leg platform undergoing construction; Ingleside, Texas.

Chevron Project Resources Company

When Chevron's Project Resources Company (PRC), which manages Chevron's major capital projects such as deep-sea and offshore oil and gas exploration and production projects, noticed an increase in quality issues, it took on a stronger role in ensuring quality by incorporating quality assurance personnel into their project team structure—a structure that previously relied more on contractor oversight and the builder's quality systems. PRC's quality inspectors are tasked with auditing and conducting surveillance of the builder's and major suppliers' quality systems and work processes. In addition, the company established a centralized quality group to provide functional support to project-level quality personnel and act as a quality advocate, so that project issues affecting quality may be elevated to a management level if necessary. From a contracting perspective, the centralized quality group created standardized contract language for quality that is used on projects managed by PRC. The project-level quality managers play a role in the contracting process in terms of tailoring the standardized quality clauses to meet the specific needs of the project and working with the builder's quality department to ensure quality requirements are well understood. PRC's team lead for quality management said that, to be effective, these quality improvement efforts required the full support of the company's leadership, a change in corporate culture, and an overall acceptance that while there is a cost to quality, the benefits are worth the costs.

Source: Chevron Project Resources Company Data.

Figure 14. Actions Taken by a Commercial Firm to Improve Quality.

Along with the formal inspections, commercial ship buyers consistently cited "roaming patrols" as central to their oversight process. Inspectors regularly patrol the shipyard, where they observe the shipyard's in- process work. All of the buyers emphasized the importance of these patrols as providing assurance that shipbuilders are adhering to their work processes even when a formal inspection is not scheduled. These patrols facilitate the early recognition of quality problems, which are typically less expensive and time consuming to correct than later in the construction process. Some officials told us that these impromptu patrols can be particularly effective in yards with more quality issues. The percentage of inspectors' time allocated to these roaming patrols can vary depending on the stage of construction—but can exceed 50 percent over the course of a ship's construction—with a greater proportion of time allocated during the earlier stages of construction when there are not as many formal inspections.

In contrast to commercial shipbuilding, where buyer and builder inspection activities are aligned through one common inspection plan, Navy shipbuilding involves layers of oversight and quality inspections that are resource intensive and not necessarily aligned with shipbuilders' inspections. In Navy shipbuilding, NAVSEA officials told us formal inspection points that relate to quality are identified in multiple ways including Navy technical documents, inspection proposals from the shipbuilder, or the ship's technical specifications. NAVSEA's engineering directorate reviews and approves the shipbuilder's quality-related test and inspection documents. Other formal test and inspection points, such as those for equipment and system installations, are contained in separate test plans that the shipbuilder develops and SUPSHIP engineering reviews. In addition to attending formal inspections, SUPSHIP's quality assurance department independently develops its own surveillance plans, and these surveillance plans are revised several times a year.[24]

Unlike inspectors for commercial ship buyers, SUPSHIP inspectors do not attend all formal inspections identified in the shipbuilding contract. SUPSHIP quality managers determine which inspections their inspectors attend—generally based on priority or potential problem areas—but it is understood that inspectors will not attend all call-outs, in part due to staffing levels at the SUPSHIP locations. NAVSEA and SUPSHIP officials noted that the unpredictability of which inspections will be attended by SUPSHIP helps ensure the shipbuilder maintains focus on all aspects of production.

SUPSHIP officials described their oversight approach in very different terms than commercial ship buyers. SUPSHIP surveillance plans set forth goals that focus on collecting numbers of observations. SUPSHIP quality assurance teams in the locations we visited stated that they generally focus most of their inspection efforts, in terms of observations conducted, on performing planned inspections identified in shipbuilding contracts and SUPSHIP surveillance plans, and less effort conducting random inspections of in-process work (similar to the roaming patrols used by commercial buyers' inspectors). The main inspection activity carried out by SUPSHIP quality inspectors, as measured by the number of observations, are product verification inspections. These are inspections of end products such as fabricated blocks, installed equipment, and completed compartments to ensure that the work conforms to the contract specifications. SUPSHIP's Gulf Coast and Bath locations reported spending 64 percent and 60 percent, respectively, of their inspection efforts on product verification inspections, while Groton and Newport News spent less time—39 percent and 35 percent, respectively.

As part of SUPSHIPs' surveillance plan for a shipyard, quality inspectors also observe work being performed by production workers—work that is in process—for compliance with the builders' procedures and technical specifications. Similar to commercial ship buyers, SUPSHIP officials stated it is important to identify, and remedy, any potential quality issues early in the construction process. Most SUPSHIP locations reported spending about 30 percent of their inspection efforts on in-process evaluations. These evaluations are planned in advance and may last a few hours to several days. SUPSHIP officials have reported that they are putting more effort into in-process inspection activities rather than just inspecting end products For example, one SUPSHIP location has established a goal of having 70 percent of their inspectors' observations related to in-process work. SUPSHIP officials also told us that their inspectors conduct some general surveillance or random inspections, but it is unclear how much of their effort is spent on these activities as they are not recorded as such.

Also in contrast to commercial shipbuilding practice, not all SUPSHIP locations use design drawings while conducting inspections. Instead, inspectors use pre-developed checklists and ship technical specifications to perform inspections. The use of design drawings during inspections helps to ensure that work is being produced in accordance with the approved design and that the builder's workers are using the correct version of the design. For example, ABS officials told us that during construction of the Littoral Combat Ship, ABS found in a number of instances that the production drawings used to build the ship were different from the approved design. In some cases the production drawings identified different sizes of pipes and flanges, or depicted piping arrangements that were not included in the approved design, but such issues generally went unnoticed by the SUPSHIP inspectors because they did not compare the design drawings to the work performed. Quality officials at two SUPSHIP locations stated that their inspectors have been trained and use the design drawings during inspections, and officials from both locations view this as an important tool for ensuring quality.[25] Officials at one of these SUPSHIP locations said they have long realized the benefits of using design drawings during the course of their quality inspectors' work and ensure the inspectors are trained to be able to read design drawings. At the other location, officials stated they just recently started having their quality inspectors use the design drawings when conducting quality inspections, which they attribute as being a good practice they observed from ABS.

Commercial Shipbuilders' Quality Management Systems Enforce Accountability for Quality Down to the Worker Level

Commercial shipbuilders create an environment of accountability for quality by implementing systems to track quality problems down to the supervisor and individual worker level. Systems used by shipbuilders included complex enterprise resource management systems that track workers assigned to specific work packages or assignments, as well as simple systems such as requiring production workers to sign their work (on the compartment itself or related paperwork) with a unique identifier. For example, one shipbuilder we visited conducts non-destructive tests on welds as dictated in the quality inspection plan developed with the ship buyer. Each test record includes information on the welder who performed the work and the outcome of the testing. This allows the shipbuilder to identify welders producing defective work. Furthermore, the shipbuilder uses the test data to rate welder performance across the yard as part of its regular performance appraisal system.

This same system allows the shipbuilder to track performance of supervisors and individual workers in terms of producing quality products and minimizing rework.

Most Navy shipbuilding contractors we met with (five of eight shipyards) reported that they have historically had difficulties identifying when and where in the production process specific quality problems occurred, as work was not always tracked at the supervisor or individual worker level. This has been a major challenge that they have been trying to address in recent years. The shipbuilders agree that quality problems are generally the result of a breakdown with the execution of their quality management plans rather than problems with the plans themselves.[26] Most of the Navy contractor shipyards we visited reported having made progress on this front, and shipbuilder quality representatives told us they have been able to improve the detection of quality problems earlier in the production process and hold front-line supervisors accountable for the quality of the work they oversee. For example, quality officials at one shipbuilder told us that their quality personnel are now focusing more on conducting in- process inspections to identify process-related problems occurring within a specific trade or work crew, as opposed to inspecting completed work products and then trying to locate the specific source of the problem within the broader production process. In another example, a shipbuilder has created quality advocates within the various tradecrafts that augment the builder's quality assurance department by assisting with day-to-day quality activities, such as conducting inspections and providing training, as well as representing production workers in yard-wide quality improvement efforts. Further, there are signs that builders are starting to hold their workforce accountable for quality issues. For example, a quality official at one shipyard we visited told us that his inspectors now routinely collect quality and defect data at the production supervisor level and develop a report of where each supervisor ranks as compared to the peer group. These reports are posted throughout the production facilities and are visible to all employees. In some rare instances, the quality official reported that his company instituted disciplinary actions for supervisors that repeatedly allowed poor quality work products to proceed to the next stage of production.

Source: Hyundai Heavy Industries (HHI).

No. 3 Dry dock at HHI, Ulsan Shipyard, South Korea.

In addition, Navy shipbuilding contractors reported that quality at some of the shipyards has been affected by high attrition rates, making it difficult to maintain a qualified workforce. One shipbuilder representative told us that his company is recruiting new technical graduates from other parts of the country to fulfill their production staff needs. Another shipbuilder representative told us that his company has had quality problems related to inexperienced labor. The need to put new staff in place to meet work demands puts pressure on the builders to expedite hiring and training. According to one shipbuilder, without a rigorous hiring and training program, under-qualified workers could be performing the work, leading to higher incidences of quality issues. Although the four international commercial shipbuilders we visited did not report having similar labor problems, we were told high turnover of the labor force is a problem at other commercial yards.

CLASSIFICATION SOCIETIES PLAY A ROLE IN COMMERCIAL AND NAVY SHIPBUILDING

In commercial shipbuilding, classification societies are an integral part of design and construction processes. All of the commercial ship buyers we met with pointed out that the role of the classification society is to ensure their ships are built in accordance with the designated classification society's rules and requirements and that it is the buyers' responsibility to ensure that shipbuilders are building ships in accordance with the buyer's requirements. Further, commercial ship buyers realize that adhering to the rules and regulations of a classification society will ensure that a new construction ship only meets the quality and safety requirements stipulated in the applicable rules.

During the contracting phase, the designated classification society's applicable rules and regulations, as well as statutory requirements (such as those pertaining to safety of life issues or marine environmental protection), are generally incorporated into the shipbuilding contract. The classification society also plays a role during the design of commercial ships, as its engineers review and approve key structural design drawings to ensure the design complies with classification society rules. These engineers also review and approve key design drawings to verify compliance with any applicable statutory requirements, if so authorized by the country where the vessel will be registered. The buyers we met with had high confidence in the classification society's engineering review of the ship design and viewed this expertise as a core competency. In some instances, commercial ship buyers noted that they contract with the classification societies for engineering services when developing and testing new technologies. For example, officials from Royal Caribbean told us they sought technical assistance from the classification society Det Norske Veritas during the development and implementation of a pilot program that tested advanced engine exhaust cleaning technologies in two of its ships.[27] Figure 15 below provides additional detail on how one commercial shipbuilding project sought technical assistance from classification societies to reduce potential risks to quality.

Classification society surveyors also have a role in monitoring ship construction; however, none of the leading buyers we met with use their designated class society as a substitute for their own construction oversight and quality assurance processes. Specifically, while classification society surveyors assess whether the ship is constructed in accordance with prescribed rules and regulations, these rules may not take into account the ship buyer's own technical specification requirements. For example, the rules do not address the quality of every structure or piece of equipment installed on the ship, such as the ship buyer's mission-related equipment, but rather only what is determined to be critical to safe operation of the ship. In addition, class societies conduct technical assessments of key parts and equipment. For a wide range of materials, parts, and equipment used in the construction of a ship, such as main engines, generators, and pumps, it is generally a class society requirement that the surveyors attend inspections and witness testing conducted at the manufacturing facility prior to the equipment being shipped to the builder's shipyard to certify that the items comply with the applicable rule requirements.

ABS No Longer Provides Classification Services on Navy Surface Combatants but Continues to Play a Role in Navy Shipbuilding

After almost 10 years of ABS participation in the development and implementation of the Naval Vessel Rules for the DDG 1000 and LCS shipbuilding programs, the Navy, in late 2011, made a decision that as a cost savings measure it would no longer seek to obtain ABS class certification for surface combatants. ABS's services were initially expected to end in June 2012, but the Navy extended ABS's involvement on construction of LCS 4 for continued marine surveying and inspection services through October 2013. The original intent of class certification was to transfer what the Navy viewed as higher-volume, lower-risk hull, mechanical, and electrical design and construction oversight work to ABS, which would allow Navy engineers to focus on higher-risk areas including mission systems and military-unique aspects of a ship such as combat systems integration. However, according to Navy engineering officials, this process was more expensive than originally envisioned.

Of the three surface combatants—LCS 1 through LCS 3—that were designed, constructed, and delivered to the Navy under the auspices of the Naval Vessel Rules, none of the ships have received ABS class certification.[28] Prior to the Navy's determination to no longer contract for ABS's services, unresolved issues precluded issuance of class certificates.[29] According to ABS officials, the classification process involved a level of discipline that the Navy found difficult to integrate into the design and construction of surface combatants, and in some instances the Navy chose to accept design drawings or approve completed production work prior to ABS completing its own review and approval process. For example, the Naval Vessel Rules require that during the design engineering phase of the shipbuilding project computer analyses determining the structural integrity of the ship are to be reviewed and approved by ABS. However, the Navy allowed the shipbuilder to commence construction and deliver the lead ship (LCS 1) and begin construction on the second ship (LCS 3) before the ship designers were able to submit a structural analysis that met the ABS requirements. The analysis identified several areas on the ship's superstructure that were under high stress and could be prone to failure. Program officials indicated they operated the ship with knowledge of the high-stress areas as a means to field test the strength of the ship. During the initial operating period of LCS 1, cracks emerged in a number of the predicted locations, requiring repair and additional strengthening of LCS 1 and structural modifications during construction of LCS 3.

Q-Max Liquefied Natural Gas Carrier

ExxonMobil and Qatar Petroleum Q-Max Liquid Natural Gas (LNG) Carriers

Project: Fabrication of Q-Max LNG carrier vessels that are designed to transport up to 80 percent more cargo than traditional LNG ships with a reduction of transportation costs of 20 to 30 percent.

Undertaking a very large ship acquisition project, Exxon Mobil and its partner, Qatar Petroleum, developed two new categories of LNG carriers. They then contracted with 3 leading shipbuilders for detailed design and construction of a number of each type of carrier including a total of 14 Q-Max ships, which are currently the world's largest class of LNG carriers. During the detailed design phase of the project, each of the shipbuilders worked with the classification societies (including ABS, Det Norske Veritas and Lloyd's Register) to conduct computer modeling and simulation analysis to ensure the structural integrity of the ships was sufficient to accommodate an anticipated 40 year service life. Also during this time, project designers and engineers from Exxon Mobil and Qatar Petroleum, along with technical experts from the classification societies, conducted their own analysis to model and test the effects LNG motion had on the structural integrity of the storage tanks under normal operating and extreme sea conditions. Modeling and testing identified tank areas that needed strengthening. These changes were incorporated into the ships' design prior to the start of construction of the lead ship without affecting the delivery schedule. Officials from Exxon Mobil stressed that this effort not only mitigated potential risks to quality, but also kept design changes to a minimum throughout the shipbuilding program.

Source: GAO analysis of ExxonMobil data.

Figure 15. Example of Classification Society Technical Assistance.

NAVSEA engineering and SUPSHIP officials indicated that the Navy's decision to cease its arrangement with ABS was also attributed to the view that the class society's role largely duplicated the work being performed by SUPSHIP and Navy engineers. These officials were confident that SUPSHIP was appropriately positioned in terms of the needed skills and resources, as their inspectors were already tasked with overseeing the builders' work to ensure compliance with the specifications. In addition, the officials informed us that NAVSEA's engineering directorate is in the process of revising the Naval Vessel Rules to take into consideration the departure of ABS from the surface combatant shipbuilding programs. SUPSHIP officials told us that a key difference between SUPSHIP's and ABS's inspection processes is the level of discretion the inspectors and surveyors have in determining whether or not work is in accordance with the prescribed rules. Specifically, it is common in commercial shipbuilding for the classification rule set to stipulate that certain work be performed to the satisfaction of the attending surveyor, which gives the classification society surveyors the flexibility to use their professional judgment as to whether or not completed work meets the intent of the rules and is fit for its purpose.[30] According to SUPSHIP officials, the Navy's revised rules will remove such language so that work performed either meets, or does not meet, the applicable technical specifications.

The Navy plans to continue work with ABS in several areas:

- NAVSEA engineering officials informed us that they will continue to contract with ABS on an as-needed basis for design and technical assistance services on new construction surface combatants as a way to augment the Navy's skill sets in specialized areas such as structural drawing reviews.
- ABS will continue to provide classification services and issue class certifications to new construction Navy ships operated by the Military Sealift Command.
- Beginning in 2008, Naval Sea Systems Command's Surface Warfare directorate contracted with ABS to perform baseline structural assessments of the condition of the surface combatant fleet when ships are dry-docked for scheduled maintenance after several years of operations.[31] Under this arrangement, scheduled to continue through fiscal year 2016, ABS is conducting marine inspections on 148 surface ships. Using the information obtained during these inspections, ABS is developing models

which identify areas of ships where there is an increased risk of failure or corrosion, which allows the Navy the ability to selectively target its maintenance activities.

More recently, in January 2013, the Navy issued a request for information and is seeking input from industry to look into the feasibility of applying commercial shipbuilding design and construction practices on its future amphibious ships. Specifically, the Navy is seeking input from industry regarding the feasibility of building an amphibious ship to ABS's commercial standards while still retaining warship capabilities. One approach identified in the request notification is to use commercial shipbuilding practices for as much of the ship as possible and apply military standards only where necessary.

CONCLUSION

The Navy pays hundreds of millions and in many cases billions of dollars for ships that warfighters rely on to perform as expected under stressing conditions. Yet it routinely accepts ships with numerous uncorrected deficiencies. Addressing these deficiencies after delivery can be costly, time consuming, and disruptive.

In recent years, Navy leadership has increased its focus on reducing what it considers to be the most serious deficiencies ("starred" deficiencies") at the time of ship delivery with some notable successes. However, the continued practice of accepting ships with a substantial number of deficiencies differs from the commercial practices we observed and can be attributed to differing interpretations of what Navy policy requires. Navy policy officials focus on provisions addressing delivery of the ship to the Navy, while program officials focus on the provisions addressing the much later point at which full financial responsibility for the ship is transferred to the operating fleet. This suggests that clarification and consistency in practice is needed. While it is understandable that there may be instances where the Navy would accept delivery of a ship with some level of uncorrected deficiencies, such actions should be the exception rather than the practice. Further, waiting until after delivery to correct known deficiencies can interfere with the activities that should be taking place during this time—ship outfitting, crew training, additional testing, and planned maintenance, as well as other class-wide upgrades. This approach is a sharp contrast to that of commercial ship buyers, who consider quality to be the focus and expect a ship that is defect-free (or nearly so) at delivery.

Navy shipbuilding entails building some of the world's most technologically advanced ships in a limited competitive environment with an industrial base that is generally reliant on the Navy to remain in business—a landscape that is much different than commercial shipbuilding. As such, any quality improvement effort must focus not only on the realities of operating in such an environment, but also on ways to incentivize the shipbuilder to produce a quality product. The Navy's Back to Basics quality improvement effort was directed at SUPSHIP and resulted—albeit in conjunction with other factors—in several positive actions that had an influence on improving quality. While the recent deliveries of the first Joint High Speed Vessel and Mobile Landing Platform represent marked improvements over previous lead ships, continued emphasis on quality and maintaining the momentum created by the

Back to Basics initiative is warranted, given that other recently delivered ships had numerous deficiencies.

While the Navy and leading commercial buyers agree that quality is the responsibility of the shipbuilder, a key difference is that the Navy makes use of cost-reimbursement and fixed-price-incentive type contracts, which assign less cost risk to the shipbuilder for quality problems. Leading commercial ship buyers have made a business decision that the risks to quality belong with the shipbuilder, and that the premium paid to transfer this risk is worth the cost of a firm-fixed-price contract. They also make greater use of how payments are structured in the shipbuilding contract to incentivize the builders to ensure timely correction of deficiencies. Even under fixed-price-incentive type contracts, when appropriate, the Navy could choose to structure certain payment provisions to incentivize shipbuilder performance and to enhance contract requirements for managing quality. For example, a standardized quality performance standard, proposed by SUPSHIP, has not been incorporated in any shipbuilding contract.

Commercial shipbuilders also attach considerable importance to having a robust oversight process. A few SUPSHIP locations have begun to make use of some oversight practices emphasized by commercial ship buyers, such as using design drawings, and these practices may be applicable to other locations as well. Further, although SUPSHIP quality inspectors conduct some level of random inspections, the Navy has not defined the role these inspections should play; and we found significant variation in their use among SUPSHIP locations.

Commercial ship buyers also place responsibility for delivering a quality ship on the site team. Responsibility for quality is more diffused in the Navy. Program offices, NAVSEA, SUPSHIP, INSURV, and others all have roles, and concerns about schedule, costs, or other strategic needs may supersede the focus on quality. While SUPSHIP's quality assurance department is closest to the work being performed, the organization has limited authority to make or exert influence on the shipbuilder and on decisions made early in the contracting process that could have a direct impact on quality. The Navy is formulating plans to establish a quality team within the NAVSEA Logistics, Maintenance and Industrial Operations Directorate that would promote, in some capacity, attention to quality assurance, but the roles, responsibilities, and authorities of the team are not yet defined. If the team is given sufficient authorities and tasked with elevating SUPSHIP quality assurance concerns throughout the acquisition process, this could present an opportunity to address some of the issues we identified in our report and emphasize the importance of quality when trade-off decisions affecting cost, schedule, and quality are at hand.

RECOMMENDATIONS FOR EXECUTIVE ACTION

To improve the construction quality of ships delivered to the Navy, we recommend that the Secretary of the Defense direct the Secretary of the Navy to take the following five actions:

1. Ensure that, when established, the NAVSEA-level quality team provides support and a direct link to directors of SUPSHIP quality assurance departments and is tasked

with raising concerns within NAVSEA about issues affecting or potentially affecting quality throughout the acquisition process.

2. Clarify Navy policy (Navy Instruction 4700.8J) by clearly identifying at what point(s) during the acquisition process contractor-responsible deficiencies are to be fully corrected and ensure the policy is followed.
3. Provide additional guidance on the quality requirements in shipbuilding contracts, including the extent to which the SUPSHIP Management Group's Quality Performance Standard for Construction of Naval Vessels should be incorporated.
4. Provide additional guidance on use of payment withholds and retentions as a means to incentivize the shipbuilding contractor to promptly correct significant or persistent deficiencies and to deliver a defect-free, or nearly defect free ship, to the Navy.
5. Assess the benefits and determine whether the following practices, in effect at some SUPSHIP locations, would be useful in detecting quality problems across all locations:
   - use of design drawings during SUPSHIP quality inspections, and
   - increased focus on random and in-process inspections compared to use of resources for other types of inspections and observations.

AGENCY COMMENTS AND OUR EVALUATION

We provided a draft of this report to DOD for comment. In its response, DOD concurred with two of our recommendations and partially concurred with three.

DOD concurred with our first recommendation, and said that it is formulating plans to develop a quality team within the Logistics, Maintenance and Industrial Operations Directorate (NAVSEA 04). This team is envisioned to provide a direct link to SUPSHIP quality assurance departments at the shipyards to enhance communication within NAVSEA on quality related issues. The quality team would also serve as a resource for identifying appropriate quality requirements in shipbuilding contracts.

DOD partially concurred with our second recommendation, to clarify Navy policy regarding acceptance of ships with quality defects. DOD said that it will continue to strive to reduce the number of open deficiencies to zero at the time the ship is delivered to the Navy. The Navy will monitor whether the trend of fewer deficiencies continues to determine if future revisions to the Navy policy instruction are needed. However, we believe the policy needs to be clarified, given the lack of consensus about whether contractor-responsible deficiencies are to be corrected before ships are delivered to the Navy. As we note in the report, addressing deficiencies post-delivery undermines the goal of constructing a defect-free or nearly defect-free ship and can also interfere with testing and crew training. Further, although trends have improved over time, recently delivered ships still had a significant number of deficiencies.

DOD partially concurred with our third recommendation, to provide guidance on quality requirements in contracts. The response cited existing FAR requirements and NAVSEA's Contract Administration Quality Assurance Program. It also stated the planned quality team within NAVSEA 04 will be a resource for consultation on appropriate contract quality assurance provisions and that this, along with the progress in reducing deficiencies, is sufficient. We believe that greater attention to contractual quality provisions is needed. SUPSHIP quality assurance staff stated that a primary reason for the development of the Quality Performance Standard for Construction of Naval Vessels was to limit inconsistencies found in shipbuilding contract quality requirements that affected SUPSHIP's ability to conduct effective quality oversight. As discussed in the report, we observed considerable variance in the specificity of contract quality provisions for different ship classes, and SUPSHIP's suggested quality performance standard has not been fully incorporated in any shipbuilding contract. Therefore we continue to believe that additional guidance on quality, including the extent to which the quality performance standard should be incorporated, would be beneficial for the Navy.

DOD concurred with our fourth recommendation, noting that, as part of its ongoing Better Buying Power initiative, it plans to better align contractor profitability with DOD goals through the use of contract incentives; actions that will assist in improving quality in shipbuilding. DOD explained that as part of this effort, the Office of the Secretary of Defense will revise and reissue guidance on the use of incentives in defense acquisition strategies. Its response noted that new tools, along with existing tools, provide sufficient opportunity for program managers and contracting officers to ensure the construction of quality ships for the Navy.

DOD also partially concurred with our fifth recommendation, to determine whether using design drawings in SUPSHIP quality inspections and increasing the focus on random and in-process inspections would be beneficial. DOD stated that use of design drawings is essential and is being used, but not necessarily by the same personnel that use checklists and other tools for inspections. It also said that random and in-process inspections are all necessary, but that the correct balance should be left to individual ship programs. We found that SUPSHIP quality assurance teams that have trained their staff to use design drawings—a commercial best practice—report benefits from this approach. Also, our work on commercial buyers' quality assurance practices consistently identified random, in-process oversight as a key tool in assuring that shipyards are following their own construction quality processes. DOD's response did not identify any additional actions the Navy plans to take to address our recommendation. We continue to believe that a SUPSHIP-wide assessment of the potential benefits of these practices could yield quality improvements.

DOD also provided technical comments that were incorporated as appropriate. These comments included questions about the deficiency data we reported for the T-AKE ship class. We reviewed individual deficiency reports (trial cards) and compared results with program office and shipbuilder reported data. We resolved anomalies and updated the data on T-AKE presented in this report.

We also provided the Navy shipbuilding contractors, commercial ship buyers, international shipbuilders, and classification societies we met with relevant excerpts of the report and incorporated their technical comments as appropriate.

Michele Mackin, Director
Acquisition and Sourcing Management

APPENDIX I: OBJECTIVES, SCOPE, AND METHODOLOGY

To identify the extent to which newly constructed Navy ships had quality problems and the actions that Navy has taken to improve quality, we reviewed the Navy's Board of Inspection and Survey reports for all Navy ships delivered from 2006 through May 2013: including *Arleigh Burke* class (DDG 51) destroyers, USS *George H. W. Bush* (CVN 77) aircraft carrier, USNS *Howard O. Lorenzen* (T-AGM 25) missile range instrumentation ship, *Lewis and Clark*-class (T-AKE) dry cargo and ammunition ships, Littoral Combat Ship (LCS)—both the *Independence and Freedom* variants, USS *Makin Island* (LHD 8) Wasp-class amphibious assault ship, USNS *Montford Point* Mobile Landing Platform, *San Antonio*-class (LPD 17) amphibious transport dock ships, USNS *Spearhead* Joint High Speed Vessel, and *Virginia*-class submarines (SSN 774). Even though the lead LPD and SSN ships were built outside of the last 8 years, we included these two vessels in our sample because the rest of the class was built within the last 8 years. We also drew from our prior work on these programs, Navy documents created to address quality issues, and various other Navy reports, such as those from the Navy's Judge Advocate General.

To determine the number and type of deficiencies for each vessel, we obtained data from the Navy's Board of Inspection and Survey's (INSURV) Material Inspection data warehouse and the Navy's Technical Support Management (TSM) system. TSM is the primary database the Navy's Supervisor of Shipbuilding, Conversion and Repair uses to track the status of new ship construction deficiencies. We analyzed these data to determine the total number of open deficiencies (1) when the ship was delivered to the Navy; (2) 120 days following ship delivery—the approximate time when final outfitting is completed prior to the ship leaving the shipyard; and (3) 1 year following ship delivery. Total deficiencies include those identified during construction, builder's trials, and acceptance trials that were not closed by the milestones listed above. Because TSM deficiency data is compiled on paper forms and manually entered into the system, we considered as being closed at delivery those deficiencies that were closed through 7 days following the date the ship was delivered to the Navy (with the exception of the T-AKE ship class), as stated on the Navy's Naval Vessel Register. To the extent feasible, we reviewed these data for completeness and for obvious inconsistency errors and compared them with paper documents that also catalogue these deficiencies. When we found obvious discrepancies while conducting our analyses, we brought them to the attention of the Navy Sea Systems Command and INSURV and worked with them to understand, correct, or omit the discrepancies. The data we collected represents the deficiencies at a particular moment in time. Further, deficiencies may be subdivided into multiple deficiencies or consolidated into a smaller number when the Navy and its shipbuilding contractors determine whether the government or the shipbuilder is responsible for correcting the respective deficiencies. For ships we reviewed, we determined that TSM deficiency data were sufficiently reliable for the purposes of this report with a few exceptions. These exceptions relate to data for T-AKE class ships. TSM data did not cover T-AKE 1 through T- AKE 6. Data for T-AKE 12 had numerous data errors and is therefore not reported on. For other T-AKE ships, we reviewed deficiency documentation (trial cards) to resolve discrepancies between TSM and T- AKE program office data.

We catalogued several hull, mechanical, and electrical issues with each ship class delivered in the last 8 years, in addition to the individual hulls that were also delivered during this period. To create this list of illustrative examples, we asked Navy officials familiar with each ship class to identify problems that occurred during the construction of the vessels. We also asked officials from INSURV to identify significant issues that affected multiple hulls within each major class. We then identified several of the issues to highlight that were illustrative of hull, mechanical, and electrical problems. Given the uniqueness of the ships' capabilities, we elected not to catalog quality issues with weapon systems or other warfighting systems. To supplement this analysis, we held discussions with or requested information from a number of Navy officials involved in Navy shipbuilding. These included the Supervisor of Shipbuilding, Conversion and Repair (SUPSHIP), Bath, Maine; Groton, Connecticut; Gulf Coast, Pascagoula, Mississippi; Newport News, Virginia; Bath Detachment, San Diego, California; Bath Detachment, Marinette, Wisconsin; and Gulf Coast Detachment, Mobile, Alabama. We also interviewed the Director of the Board of Inspection and Survey and officials in the Naval Sea Systems Command Management Group for SUPSHIP; Naval Sea Systems Command Contracting Directorate; Naval Sea Systems Command Engineering Directorate; Naval Sea Systems Command Surface Warfare Directorate; Naval Sea Systems Command Nuclear Propulsion; Program Executive Offices for Carriers; Program Executive Office for Submarines, and the Littoral Combat Ship; representatives from acquisition program offices including PMS 317 (LPD 17), PMS 377 (LHD 8 and LHA 6), PMS 385 (Joint High Speed Vessel and Mobile Landing Platform), PMS 400 D (DDG 51), and, PMS 501 (LCS); Norfolk Ship Support Activity—Regional Maintenance Center; Southwest Regional Maintenance Center; Commander, Operational Test and Evaluation Force; Military Sealift Command; Fleet Forces Command; and the Defense Contract Management Agency.

We also visited eight U.S. Navy contractor shipyards that build some of the larger Navy vessels and met with representatives from the contractor that owns each shipyard, including Austal USA in Mobile, Alabama; General Dynamics Electric Boat Corporation in Groton, Connecticut and Quonset Point, Rhode Island; General Dynamics NASSCO in San Diego, California, and General Dynamics Bath Iron Works in Bath, Maine; Huntington Ingalls Industries Ingalls Shipbuilding in Pascagoula, Mississippi and Huntington Ingalls Industries Newport News Shipbuilding in Newport News, Virginia; Marinette Marine Corporation in Marinette, Wisconsin; and, V.T. Halter Marine at its shipyard in Pascagoula, Mississippi. In addition, we observed the underway portion of the acceptance trial for T-AKE 13 in San Diego, California.

Lastly, we reviewed the Navy's efforts to improve ship quality by reviewing key memos and documents outlining the Back to Basics program and meeting with the officials responsible for implementing these efforts—most of whom were in the Navy's Supervisor of Shipbuilding, Conversion and Repair commands. We also reviewed selected parts of 11 fixed-priced-incentive and firm-fixed priced Navy shipbuilding contracts, such as clauses pertaining to quality requirements and ship delivery, for *San Antonio* class (LPD 17) amphibious transport dock ships (one contract); *America* class (LHA 6 and LHA 7) amphibious assault ship (two contracts); *Arleigh Burke* class (DDG 51) destroyer (one contract); Joint High Speed Vessel (one contract); Littoral Combat Ship (LCS)— *Freedom* and *Independence* variants (two contracts); *Lewis and Clark* class (T-AKE) dry cargo and ammunition ship (one contract); USNS *Howard O. Lorenzen* (T-AGM 25) missile instrumentation ship (one contract) ; USS *Makin Island* (LHD 8) amphibious assault ship (one contract); and the Mobile Landing Platform (one contract).

To assess key practices used by commercial ship buyers and shipbuilders, we interviewed and met with leading commercial companies from the cruise, oil and gas, and commercial shipping industries, including Carnival Corporation; Chevron Corporation; Ensco plc; ExxonMobil; A.P. Moller-Maersk A/S; Noble Corporation; Norwegian Cruise Lines; Royal Caribbean Cruises, Ltd.; and Seadrill, Ltd. We identified leading companies by analyzing such indicators as annual sales, number of vessels owned and procured, and total market share. Our methodology drew from our previous shipbuilding best practices work that identified the commercial shipbuilding industries that support cruise, oil and gas, and commercial shipping sectors as being most similar to Navy shipbuilding. Cruise ships are more costly and complex than other types of commercial ships, densely packed, and require significant outfitting, making them somewhat similar to military ships. Additionally, cruise ship buyers often include innovations or design changes in their ships and start new classes of ships regularly in order to maximize passenger satisfaction; approaches that allowed us to examine quality oversight practices on recent lead ship programs and the outcomes of specific commercial practices. The cruise line companies we met with are leaders in their industry as identified in our previous work and based on operating revenue or fleet size. We met with ship buyers from the oil and gas industry because drill ships, floating production storage and offloading (FPSO) vessels and offshore oil platforms are complex, dense structures. Furthermore, FPSOs, essentially floating refineries, are often one of a kind, costing well over $1 billion. The oil and gas companies we met with are leaders in their industry as identified in our previous work as well as our assessment of top operators of drilling vessels. Similarly, we met with Maersk Line Limited and Maersk Drilling, two business units within A.P. Moller-Maersk A/S, because the company was identified as an industry leader in our prior work and remains one of the largest shipping companies in the world. For example, Maersk Line Limited acquires many ships: in 2012 the company took delivery of 19 new ships. For each commercial ship buyer, we requested deficiency data on one or more new construction ships they had acquired. With the exception of one FPSO, all of these ships were delivered to the buyer in 2012 or 2013. We assessed the reliability of this data by obtaining information on the systems that stored the data and interviewing ship buyer and shipbuilder representatives knowledgeable about the data. We determined that the data were sufficiently reliable for the purposes of this report. We also identified common processes and tools used by these ship buyers and shipbuilders to ensure the expected level of quality.

To determine the extent to which Navy quality assurance processes use common commercial practices, we reviewed data and information obtained from the Navy and its shipbuilding contractors, as well as from the leading commercial ship buyers and shipbuilders. We also reviewed relevant payment and retention clauses for 11 fixed-priced Navy shipbuilding contracts. During our meetings with SUPSHIP and program, engineering, and contracting officials from the Naval Sea Systems Command directorates, we inquired about the Navy's quality assurance practices. During site visits to eight U.S. private shipyards that build Navy ships, we discussed with shipyard representatives their quality assurance processes and the steps taken to ensure their ships meet the Navy's quality expectations. During our site visits, we collected documentation related to the shipyard's quality assurance activities, including quality policies, plans, and procedures, and we also observed quality assurance inspection activities when possible.

We also met with officials from international commercial shipyards that are responsible for building a variety of complex ships, including Meyer Werft (Germany) and STX Finland

(Finland), which both build cruise ships; and Hyundai Heavy Industries and Daewoo Shipbuilding and Marine Engineering (South Korea), which build commercial ships, including containerships, liquefied natural gas carriers, drill ships, FPSOs, and oil tankers. We identified these shipbuilders as producers of high- quality vessels through a combination of our previous work and recommendations from shipbuilding experts and the ship buyers that participated in our review. At the shipyards, we met with ship buyers' representatives who were responsible for overseeing the construction of the ships and monitoring the construction schedule. Where possible, we observed quality assurance activities at commercial shipbuilders or viewed systems related to ensuring quality, such as quality database systems. We collected documentation of quality assurance activities, such as quality policies and inspection plans, where available. We met with representatives from three classification societies, including the American Bureau of Shipping (ABS), Det Norske Veritas, and Lloyd's Register regarding their roles in commercial ship construction. We also met with representatives from Aker Shipyard in Philadelphia, Pennsylvania, to discuss how a U.S. shipbuilder that solely builds commercial vessels approaches quality assurance.

To better understand the role of classification societies in Navy and commercial shipbuilding, we met with engineering and marine surveying representatives from ABS to obtain an overview of how they conduct their work. We held discussions on the role of classification in Navy shipbuilding with Navy shipbuilding contractors, including Austal USA, Mobile, Alabama; General Dynamics Bath Iron Works, Bath, Maine; General Dynamics NASSCO, San Diego, California; Marinette Marine Corporation, Marinette, Wisconsin; and V.T. Halter Marine, Pascagoula, Mississippi. To learn the extent to which the Navy's approach to new ship construction oversight is similar to or different than the marine surveying services provided by ABS during construction of Navy ships, we met with officials from SUPSHIP locations at Bath, Maine; Bath Detachment, San Diego, California; Bath Detachment, Marinette, Wisconsin; Gulf Coast, Pascagoula, Mississippi; and Mobile, Alabama. We also held discussions on this matter with officials from the Naval Sea Systems Command SUPSHIP Management Group; Military Sealift Command; Naval Sea Systems Command Engineering Directorate; Naval Sea Systems Command Surface Warfare Directorate; Program Executive Office Littoral Combat Ship; and representatives from acquisition program offices including PMS 385 (Joint High Speed Vessel and Mobile Landing Platform). We reviewed the classification rule set developed by the Navy and ABS (the Naval Vessel Rules), as well as other classification rule sets pertaining to Navy and or commercial vessels, such as High-Speed Naval Craft rules and the Steel Vessel rules. In addition, we reviewed the findings and observations of ABS's marine surveyors for the Littoral Combat Ship which is being built in accordance with the Naval Vessel Rules, as well as the Joint High Speed Vessel, which is built in accordance with ABS's Naval High Speed Naval Craft Guide. We also spoke with representatives from other classification societies, including Det Norske Veritas and Lloyd's Register, to discuss their approach to classification of commercial and navy vessels. When meeting with commercial ship buyers and shipbuilders, we also discussed the ship classification process and the role of classification societies in shipbuilding.

We conducted this performance audit from March 2012 to November 2013 in accordance with generally accepted government auditing standards. Those standards require that we plan and perform the audit to obtain sufficient, appropriate evidence to provide a reasonable basis for our findings and conclusions based on our audit objectives. We believe that the evidence obtained provides a reasonable basis for our findings and conclusions based on our audit objectives.

APPENDIX II: STAGES OF SHIPBUILDING

There are four primary phases in shipbuilding: pre-contracting, contract award, design and planning, and construction, with each phase building upon the work completed in earlier stages. Within each phase, a number of key events have an influence on the overall quality of the ship. In addition, within Navy shipbuilding, additional key activities take place following ship delivery. Table 2 describes some of the more significant events occurring throughout the shipbuilding process.

Table 2. Key Events Occurring During Navy and Commercial Shipbuilding

Stage	Key event	Description
Pre-contracting activities Contract award	Concept refinement	Ship buyer determines necessary requirements and desired capabilities, develops an acquisition strategy.
	Early-stage design	Ship buyer refines its operational and performance requirements into specifications that will be included in the shipbuilding contract.
	Contract award and negotiation	Ship buyer selects and enters into a shipbuilding contract with the chosen shipbuilder(s).The contract includes the ship's specification, which details how the shipbuilder will build the ship and meet the buyer's requirements.
Design and planning	Detailed engineering design	Ship designer develops all aspects of the ship's structure and routing of major distributive systems, such as electrical or piping, throughout the ship. A three-dimension (3D) computer-aided-design model is often generated, along with completion of any computer modeling or simulation analyses, such as those to test the structural integrity of the ship design throughout its service life or under certain sea conditions.
	Pre-construction and planning activities	Shipbuilder plans production flow and develops two-dimensional paper drawings that, once approved by the ship buyer, will be used by shipyard workers to build the ship. Ship buyer, shipbuilder, and classification society (if applicable) collectively determine quality-related test and inspection points during ship construction.
Construction	Steel cutting/block fabrication	Ship fabrication begins as large steel or aluminum plates are cut and welded to form the basic building units for a ship called "blocks." Blocks comprise compartments, which include accommodation space, engine room, and storage areas.
	Assembly and outfitting of blocks	Upon completion of a block, piping, brackets for machinery or cabling, and ladders, among other things, are installed. Installing these items at this stage is preferable because access to spaces is not limited by doors or other machinery, requiring less time and effort than at later stages of construction.

Stage	Key event	Description
	Keel laying and block erection	Blocks are welded to form larger sections, referred to as grand blocks, which comprise the ship's structure. The shipbuilder then assembles and welds grand blocks and blocks in the drydock to form the keel. Machinery, engines, propeller shafts and other large items are also installed during this stage.
	Launch	Once the ship is watertight, the drydock is flooded and the ship is towed to a docking area where final outfitting of machinery and equipment occur.[a]
	System testing and commissioning	Parts, materials, and machinery, such as engines, pumps, and associated control instrumentation used in the ship, are generally tested by the manufacturer (factory acceptance test) to ensure quality standards, technical specifications, and performance requirements are met. Installation and connection of these components create subsystems. The shipbuilder and ship buyer ensure the subsystems and systems are installed in accordance to the ship's specifications and conduct tests to ensure systems are operating as intended and meet performance requirements.
	Sea trials	Once the shipbuilder is satisfied that the ship is seaworthy and meets the buyer's requirements, the ship buyer's representatives, and if applicable the classification society's surveyors, are brought onboard and the ship embarks on a series of dockside and at-sea tests where the overall quality and performance of the ship is evaluated against the contractually required specifications. Sea trials provide early verification of the buyer's requirements and allow time for any corrective actions that may be required to meet the buyer's requirements prior to ship delivery. Navy shipbuilding programs generally conduct two sets of sea trials—builder's trials and acceptance trials. Builder's trials test the vessel's propulsion, communications, navigation and mission systems, as well as all related support systems. Following the successful completion of builder's trials, acceptance trials are conducted by the Navy's Board of Inspection and Survey (INSURV).
	Delivery/Acceptance	Ship buyer takes custody and assumes ownership of the vessel. In the commercial world, the ship is complete and commences operations. In Navy shipbuilding, a Material Inspection and Receiving Report (Form DD 250) is prepared, representing the official transfer of custody and ownership to the Navy. Any unresolved deficiencies or remaining work items are segregated by the entity that is responsible for completion of the work (Navy or shipbuilder) and identified on this document.
Post-delivery activities specific to the Navy	Final outfitting	Crew boards the ship and begins training; and mission systems are installed.[b]
	Post-delivery tests and trials	Operational tests are conducted on the ships combat and mission critical systems.
	Final contract trials	INSURV conducts a second round of sea trials just prior to the expiration of the ship's guarantee period.[c]

Table 2. (Continued)

Stage	Key event	Description
	Post Shakedown Availability	Planned maintenance period prior to the maiden voyage where work is performed to install class-wide upgrades or ship improvements, perform maintenance, and correct new or previously identified construction deficiencies. Usually performed using a different contract than shipbuilding contract.
	Shipbuilding and Conversion, Navy Obligation and Work Limiting Date	The official date where full responsibility for funding the ship's operation and maintenance is transferred from the acquisition command to the operational fleet.

Source: GAO analysis of Navy and industry provided data.

[a] The level of outfitting completed prior to launch varies by shipbuilder and ship type, but is predetermined according to the builder's production plan. Shipbuilders generally agree that launching a ship having a lower level of outfitting completed than what was planned can increase the costs to complete the work.

[b] On nuclear-powered Navy ships, the ship's crew begins boarding and training prior to ship delivery.

[c] The guarantee period is the time after delivery where the shipbuilder is responsible for correcting any defects or deficiencies in accordance to the terms and conditions of the contract.

APPENDIX III: LEWIS AND CLARK DRY CARGO AND AMMUNITION SHIP CLASS (T-AKE)

The *Lewis and Clark* class of dry cargo and ammunition ships (T-AKE 1) consists of 14 ships which have been delivered to the Navy. The first ship was delivered in 2006 and the final ship was delivered in 2012. We analyzed data provided by the Supervisor of Shipbuilding, Conversion, and Repair (SUPSHIP) to determine the number of open deficiencies at delivery, 120 days after delivery, and one year after delivery. Our analysis of the available data found that recently delivered T-AKEs had noticeably fewer open deficiencies at delivery compared to other ship classes. Unlike other classes in our review, the Navy was responsible for the majority of these deficiencies.

Table 3. Open Deficiencies on T-AKE Class Ships at Various Points in Time after Delivery

Ship (delivery date)	Importance[b]	Delivery[a]		120 days after delivery		365 days after delivery	
		Total deficiencies[c]	Percentage of contractor-responsible deficiencies[d]	Total deficiencies	Percentage of contractor-responsible deficiencies	Total deficiencies	Percentage of contractor-responsible deficiencies
T-AKE 7 (March 2009)	Part 1	22	5	5	0	1	0
	Part 2	154	1	44	2	3	0
	Part 3	3	0	2	0	0	0

Ship (delivery date)	Importance[b]	Delivery[a]		120 days after delivery		365 days after delivery	
		Total deficiencies[c]	Percentage of contractor-responsible deficiencies[d]	Total deficiencies	Percentage of contractor-responsible deficiencies	Total deficiencies	Percentage of contractor-responsible deficiencies
T-AKE 8 (September 2009)	Part 1	5	20	0	0	0	0
	Part 2	40	18	18	6	2	50
	Part 3	1	0	0	33	0	0
T-AKE 9 2010)	Part 1	29	8	4	25	1	100
	Part 2	136	2	14	7	1	100
	Part 3	5	0	0	0	0	0
T-AKE 10 (July 2010)	Part 1	5	40	5	40	4	50
	Part 2	39	8	21	14	10	10
	Part 3	0	0	0	0	0	0
T-AKE 11 (February 2011)	Part 1	6	0	1	0	0	0
	Part 2	17	12	5	20	2	50
	Part 3	0	0	0	0	0	0
T-AKE 13 (April 2012)	Part 1	3	0	2	0	1	0
	Part 2	32	4	24	4	4	0
	Part 3	0	0	0	0	0	0
T-AKE 14 (October 2012)	Part 1	3	0	0	0	-	-
	Part 2	21	0	1	0	-	-
	Part 3	0	0	0	0	-	-

Source: GAO analysis of Navy data as of March 2013.

Notes: (1) TSM relies on user-entered information, which can be unreliable. T-AKE 12's TSM data did not pass our data reliability standards because 98 percent of the deficiencies were opened and closed in the system on the same day—a date after the ship had been delivered to the Navy. (2) Deficiencies with missing Importance designations are not included in this table.

[a] The table counts deficiencies from the Navy's Technical Support Management (TSM) system which were closed after the official delivery date listed in the Naval Vessel Register, www.nvr.navy.mil, including those deficiencies where it was subsequently determined no further corrective action would be taken. In some instances, similar types of deficiencies were consolidated into a single deficiency prior to ship delivery. In other instances, uncorrected deficiencies were closed and transferred to final contract trials deficiencies (trial cards).

[b] Deficiencies are numbered by their significance and order of importance as Part 1, Part 2, and Part 3. The Navy's Board of Inspection and Survey (INSURV) defines a Part 1 deficiency is an important deficiency which is likely to cause the ship to be unseaworthy, substantially reduce the ability of the ship to carry out an assigned mission, or cause serious injury to personnel or serious damage to important material or equipment. According to INSURV, Part 2 deficiencies are less significant or do not meet the criteria for Part 1 deficiencies, but should be corrected to restore the ship to required specifications. INSURV classifies Part 3 deficiencies as those that will require either major alterations to correct (design related) or modifications that are too costly to effect during the life cycle of the ship.

[c] The Navy's Board of Inspection and Survey (INSURV) documents all deficiencies which require corrective action to bring the material condition of the ship to required specifications. Deficiencies may have been identified during INSURV or builder's trials.

[d] Percentage is rounded to the nearest whole percent. Navy program managers and SUPSHIP may assign responsibility for correcting a deficiency to the contractor if the contractor, sub-contractors, or vendors do not meet the requirements of the shipbuilding contract. Where data were available, we determined the percent of Part 1, Part 2, Part 3, and total deficiencies designated as contractor-responsible versus government-responsible. In some cases, the responsible party for correcting a deficiency may alternate between the Navy and the contractor based on additional investigations into the issues.

APPENDIX IV: *SAN ANTONIO* AMPHIBIOUS TRANSPORT DOCK SHIP CLASS (LPD 17)

The *San Antonio* class of amphibious transport docks (LPD 17) consists of eight ships which have been delivered to the Navy. The first ship was delivered in 2005. As of September 2013, the Navy has three ships under construction. The *San Antonio* class has generally seen a decline in the number of open deficiencies at delivery, yet each ship still has a large number of open deficiencies.

Table 4. Open Deficiencies on LPD 17 Class Ships at Various Points in Time after Delivery

Ship (delivery date)	Importance[b]	Delivery[a]		120 days after delivery		365 days after delivery	
		Total deficiencies[c]	Percentage of contractor-responsible deficiencies[d]	Total deficiencies	Percentage of contractor-responsible deficiencies	Total deficiencies	Percentage of contractor-responsible deficiencies
LPD 21 August 2009)	Part 1	234	58	75	28	23	26
	Part 2	6,078	79	1,385	49	465	57
	Part 3	13	23	0	0	0	0
LPD 22 (December 2011)	Part 1	156	25	102	5	47	2
	Part 2	4,783	69	1,074	49	453	57
	Part 3	24	63	1	100	0	0
LPD 23 (September 2012)	Part 1	88	25	55	18	-	-
	Part 2	1,270	42	685	51	-	-
	Part 3	45	2	37	0	-	-
LPD 24 (December 2012)	Part 1	86	48	36	25	-	-
	Part 2	1,891	58	667	48	-	-
	Part 3	28	25	15	13	-	-

Source: GAO analysis of Navy data as of March 2013.

Notes: (1) Deficiencies with missing Importance designations are not included in this table.

[a] For the purposes of this report, the table counts deficiencies from the Navy's Technical Support Management (TSM) system which were closed seven days or more after the official delivery date listed in the Naval Vessel Register, www.nvr.navy.mil. Deficiencies closed after delivery but before the seventh day are not included.

[b] Deficiencies are numbered by their significance and order of importance as Part 1, Part 2, and Part 3. INSURV defines a Part 1 deficiency is an important deficiency which is likely to cause the ship to be unseaworthy, substantially reduce the ability of the ship to carry out an assigned mission, or cause serious injury to personnel or serious damage to important material or equipment. According to INSURV, Part 2 deficiencies are less significant or do not meet the criteria for Part 1 deficiencies, but should be corrected to restore the ship to required specifications. INSURV classifies Part 3 deficiencies as those that will require either major alterations to correct (design-related) or modifications that are too costly to effect during the life cycle of the ship.

[c] The Navy's Board of Inspection and Survey (INSURV) documents all deficiencies which require corrective action to bring the material condition of the ship to required specifications. Deficiencies may have been identified during INSURV or builder's trials.

[d] Percentage is rounded to the nearest whole percent. Navy program managers and SUPSHIP may assign responsibility for correcting a deficiency to the contractor if the contractor, sub-contractors, or vendors do not meet the requirements of the shipbuilding contract. Where data were available, we determined the percent of Part 1, Part 2, Part 3, and total deficiencies designated as contractor-responsible versus government-responsible. In some cases, the responsible party for correcting a deficiency may alternate between the Navy and the contractor based on additional investigations into the issues.

We analyzed data provided by the Supervisor of Shipbuilding, Conversion, and Repair (SUPSHIP) to determine the number of open deficiencies at delivery, 120 days after delivery, and one year after delivery. Our analysis found that recently delivered ships had thousands of open deficiencies, ranging from 1,403 on LPD 23 to 6,325 on LPD 21. The majority of the deficiencies open at delivery were the responsibility of the contractor with the exception of LPD 23 where only 40 percent of the open deficiencies were the contractor's responsibility. The data below indicates that many of these deficiencies are being closed after the ships are delivered to the Navy and are being outfitted.

APPENDIX V: LITTORAL COMBAT SHIP (LCS) CLASS

The Littoral Combat Ship (LCS) class consists of 3 ships which had been delivered to the Navy at the time of our review. The first hull was delivered in 2008. Currently, the Navy has seven ships under construction and has received funding to construct six ships. The Navy has an additional eight ships under contract which are not yet funded. The LCS class consists of two different seaframe designs, the LCS 1 design (*Freedom* variant) and the LCS 2 design (*Independence* variant).

We analyzed data provided by the Supervisor of Shipbuilding, Conversion, and Repair (SUPSHIP) to determine the number of open deficiencies at delivery, 120 days after delivery, and one year after delivery. Our analysis found that both designs of the LCS class were delivered with a large number of open deficiencies at delivery. The majority of these deficiencies were the responsibility of the contractor. Our analysis found that over half of these deficiencies were closed after the ships were delivered to the Navy and were being outfitted.

Table 5. Open Deficiencies on LCS Class Ships at Various Points in Time after Delivery

Ship (delivery date)	Importance[b]	Delivery[a]		120 days after delivery		365 days after delivery	
		Total deficiencies[c]	Percentage of contractor-responsible deficiencies[d]	Total deficiencies	Percentage of contractor-responsible deficiencies	Total deficiencies	Percentage of contractor-responsible deficiencies
LCS 1 (September 2008)	Part 1	151	72	86	62	30	37
	Part 2	1,994	81	988	76	175	46
	Part 3	109	82	56	77	11	36
LCS 2 (December 2009)	Part 1	547	81	256	71	131	64
	Part 2	3,715	71	2,079	60	819	56
	Part 3	954	82	510	74	213	62
LCS 3 (June 2012)	Part 1	103	57	40	63	-	-
	Part 2	1,178	75	482	65	-	-
	Part 3	18	50	5	80	-	-

Source: GAO analysis of Navy data as of March 2013.

Notes: (1) Odd numbered LCS class ships (*Freedom* variant) are built at a Navy contractor shipyard in Marinette, Wisconsin, while even numbered ships (*Independence* variant) are built at a Navy contractor shipyard in Mobile, Alabama. (2) Deficiencies with missing Importance designations are not included in this table.

[a] For the purposes of this report, the table counts deficiencies from the Navy's Technical Support Management (TSM) system which were closed seven days or more after the official delivery date listed in the Naval Vessel Register, www.nvr.navy.mil. Deficiencies closed after delivery but before the seventh day are not included.

[b] Deficiencies are numbered by their significance and order of importance as Part 1, Part 2, and Part 3. INSURV defines a Part 1 deficiency is an important deficiency which is likely to cause the ship to be unseaworthy, substantially reduce the ability of the ship to carry out an assigned mission, or cause serious injury to personnel or serious damage to important material or equipment. According to INSURV, Part 2 deficiencies are less significant or do not meet the criteria for Part 1 deficiencies, but should be corrected to restore the ship to required specifications. INSURV classifies Part 3 deficiencies as those that will require either major alterations to correct (design related) or modifications that are too costly to effect during the life cycle of the ship.

[c] The Navy's Board of Inspection and Survey (INSURV) documents all deficiencies which require corrective action to bring the material condition of the ship to required specifications. Deficiencies may have been identified during INSURV or builder's trials.

[d] Percentage is rounded to the nearest whole percent. Navy program managers and SUPSHIP may assign responsibility for correcting a deficiency to the contractor if the contractor, sub-contractors, or vendors do not meet the requirements of the shipbuilding contract. Where data were available, we determined the percent of Part 1, Part 2, Part 3, and total deficiencies designated as contractor-responsible versus government-responsible. In some cases, the responsible party for correcting a deficiency may alternate between the Navy and the contractor based on additional investigations into the issues.

APPENDIX VI: *ARLEIGH BURKE* GUIDED MISSILE DESTROYER SHIP CLASS (DDG 51)

The *Arleigh Burke* class of guided missile destroyers (DDG 51) consists of 62 ships which had been delivered to the Navy at the time of our review. The first ship was delivered in 1991. Currently, the Navy has received funding to construct an additional four ships. Despite being a well established program, the *Arleigh Burke* class continues to have a large number of open deficiencies at various points in time.

We analyzed data provided by the Supervisor of Shipbuilding, Conversion, and Repair (SUPSHIP) to determine the number of open deficiencies at delivery, 120 days after delivery, and one year after delivery. Our analysis found that recently delivered ships had a large number of open deficiencies, ranging from a low of 333 open deficiencies on DDG 110 to a high of 4,385 open deficiencies on DDG 109. The majority of the deficiencies open at delivery were the responsibility of the contractor. The data below indicates that many of these deficiencies were being closed after the ships had been delivered to the Navy and were being outfitted.

Table 6. Open Deficiencies on DDG 51 Class Ships at Various Points in Time after Delivery

Ship (delivery date)	Importance[b]	Delivery[a]		120 days after delivery		365 days after delivery	
		Total deficiencies[c]	Percentage of contractor-responsible deficiencies[d]	Total deficiencies	Percentage of contractor-responsible deficiencies	Total deficiencies	Percentage of contractor-responsible deficiencies
DDG 108 (July 2009)	Part 1	38	47	17	0	6	0
	Part 2	1,186	72	100	11	52	4
	Part 3	1,981	90	94	24	23	35
DDG 105 (August 2009)	Part 1	96	70	15	27	8	50
	Part 2	1,453	73	116	33	61	61
	Part 3	487	92	22	59	19	68
DDG 109 (June 2010)	Part 1	52	71	10	20	3	67
	Part 2	2,483	81	172	36	59	8
	Part 3	1,850	90	122	63	29	14
DDG 107 (July 2010)	Part 1	16	56	7	14	4	25
	Part 2	672	63	61	38	31	29
	Part 3	235	86	7	71	3	100
DDG 110 (February 2011)	Part 1	12	25	8	13	4	25
	Part 2	285	59	55	33	36	36
	Part 3	35	66	1	0	0	0
DDG 111 (April 2011)	Part 1	17	29	5	0	2	0
	Part 2	2,057	74	279	36	140	15
	Part 3	1,896	92	152	63	41	5
DDG 112 (May 2012)	Part 1	20	20	5	0	-	-
	Part 2	1,085	55	159	20	-	-
	Part 3	0	0	0	0	-	-

Source: GAO analysis of Navy data as of March 2013.

Notes: (1) DDG 51 class ships are built at two Navy contractor shipyards in Bath, Maine and Pascagoula, Mississippi. (2) The ships are presented in chronological based on the date of delivery to the Navy. (3) Deficiencies with missing Importance designations are not included in this table.

[a] For the purposes of this report, the table counts deficiencies from the Navy's Technical Support Management (TSM) system which were closed seven days or more after the official delivery date listed in the Naval Vessel Register, www.nvr.navy.mil. Deficiencies closed after delivery but before the seventh day are not included.

[b] Deficiencies are numbered by their significance and order of importance as Part 1, Part 2, and Part 3. INSURV defines a Part 1 deficiency is an important deficiency which is likely to cause the ship to be unseaworthy, substantially reduce the ability of the ship to carry out an assigned mission, or cause serious injury to personnel or serious damage to important material or equipment. According to INSURV, Part 2 deficiencies are less significant or do not meet the criteria for Part 1 deficiencies, but should be corrected to restore the ship to required specifications. INSURV classifies Part 3 deficiencies as those that will require either major alterations to correct (design related) or modifications that are too costly to effect during the life cycle of the ship.

[c] The Navy's Board of Inspection and Survey (INSURV) documents all deficiencies which require corrective action to bring the material condition of the ship to required specifications. Deficiencies may have been identified during INSURV or builder's trials.

[d] Percentage is rounded to the nearest whole percent. Navy program managers and SUPSHIP may assign responsibility for correcting a deficiency to the contractor if the contractor, sub-contractors, or vendors do not meet the requirements of the shipbuilding contract. Where data were available, we determined the percent of Part 1, Part 2, Part 3, and total deficiencies designated as contractor- responsible versus government-responsible. In some cases, the responsible party for correcting a deficiency may alternate between the Navy and the contractor based on additional investigations into the issues.

APPENDIX VII: SELECTED QUALITY ISSUES IN NAVY SHIP CLASSES

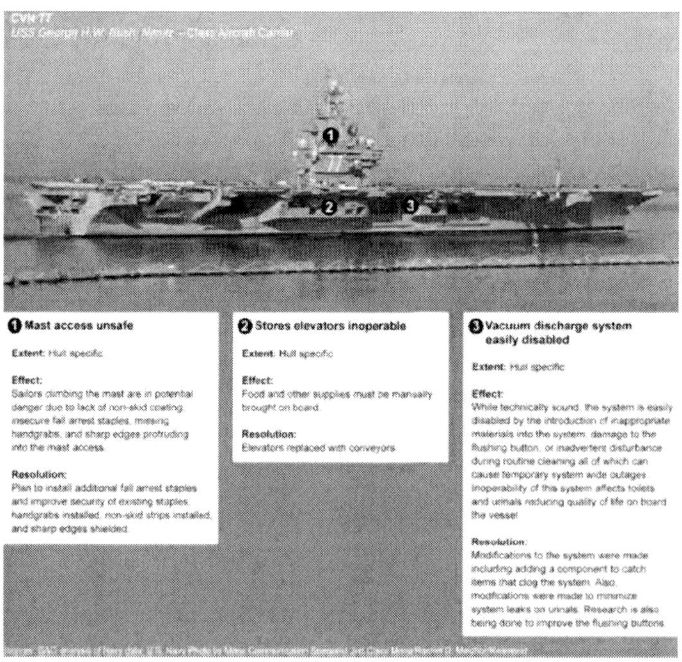

Figure 16. Selected Quality Issues on USS George H.W. Bush (CVN 77).

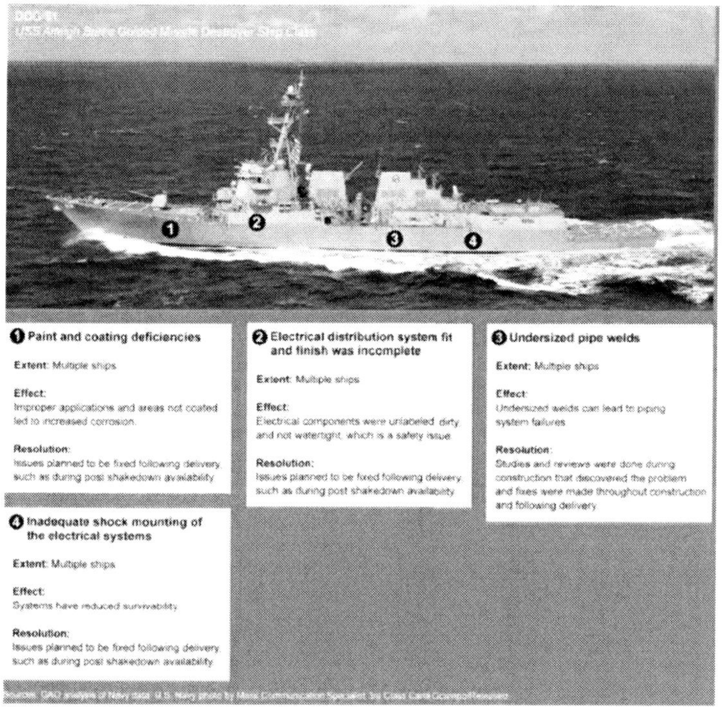

Figure 17. Selected Quality Issues on DDG 51 Class Ships.

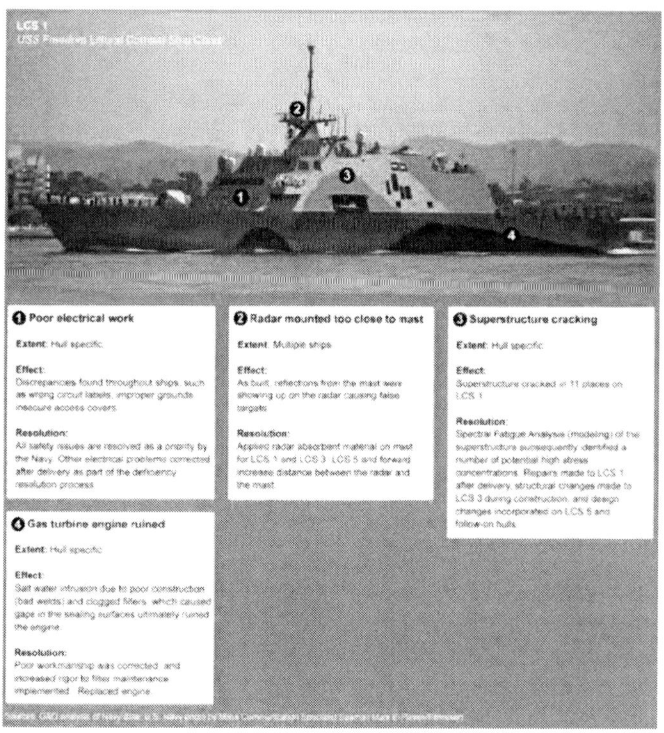

Figure 18. Selected Quality Issues on USS Freedom (LCS 1).

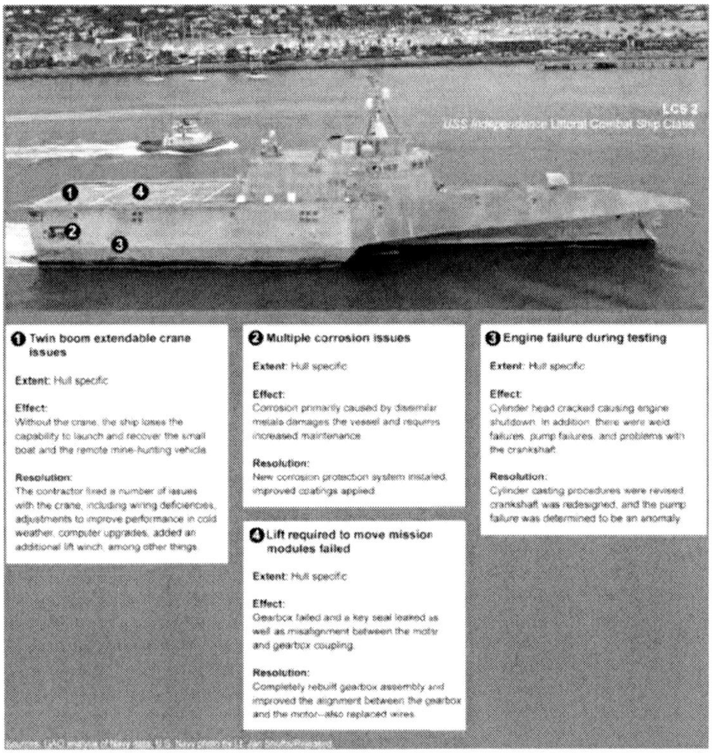

Figure 19. Selected Quality Issues on USS In dependence (LCS 2).

Figure 20. Selected Quality Issues on USS Makin Island (LHD 8).

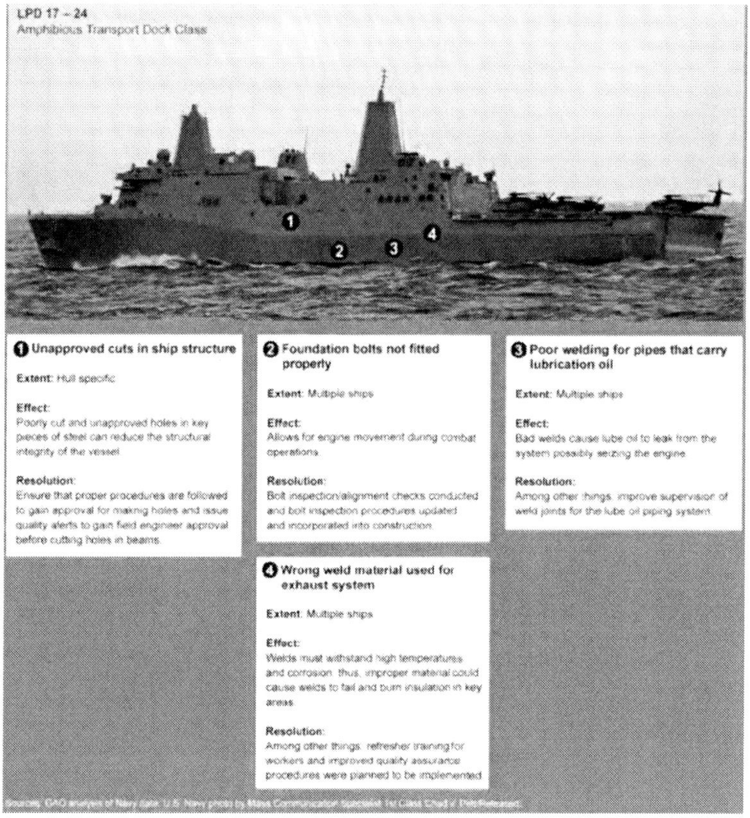

Figure 21. Selected Quality Issues on LDP 17 Class Ships.

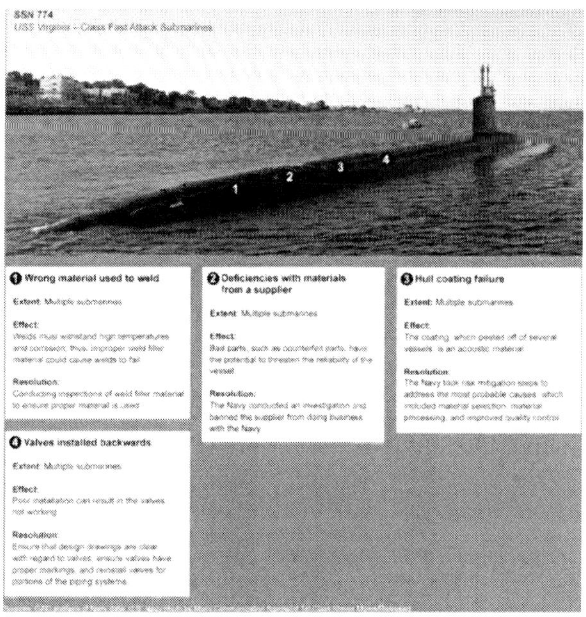

Figure 22. Selected Quality Issues on SSN 774 Class Submarines.

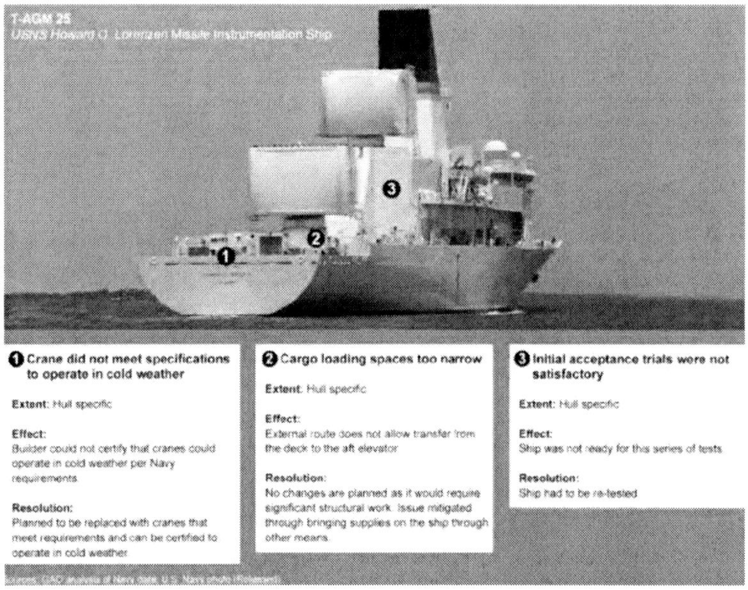

Figure 23. Selected Quality Issues on USNS Howard O. Lorenzen (T-AGM 25).

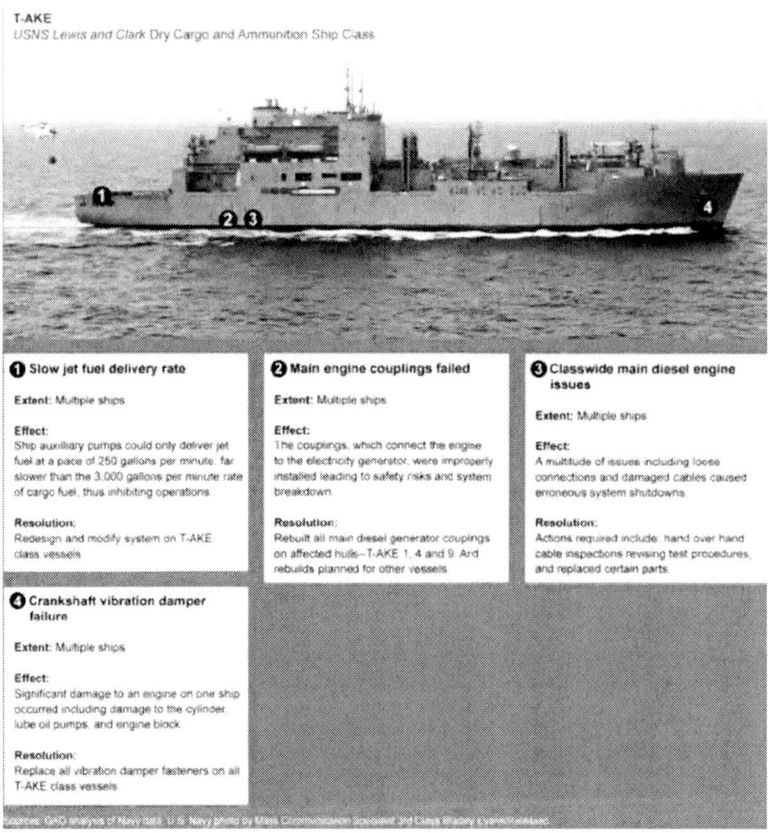

Figure 24. Selected Quality Issues on T-AKE Class Ships.

APPENDIX VIII: CHARACTERISTICS OF SELECT COMMERCIAL SHIPS AND OFFSHORE STRUCTURES

Each of the ship buyers we met with acquires large, expensive, and technologically advanced vessels. Below are illustrative examples of ships and offshore structures acquired by the firms we met with, and some of the key characteristics associated with each of the respective vessels.

Table 7. Key Characteristics of Selected Commercial Ships and Offshore Structures

Ship	Builder	Approximate cost (U.S. Dollars)	Length (feet)	Displacement (tons)	Notable characteristics
Celebrity Cruises, Inc. (subsidiary of Royal Caribbean Cruises, LTD.) *Reflection* – cruise ship (delivered October 2012)	Meyer Werft, Germany	$750 million	1,047 feet	61,609 tons displacement	The fifth and last ship of the *Solstice* ship class, completing a $3.7 billion shipbuilding program. The ship has a crew of 1,253, can accommodate 3,223 passengers, and incorporated advanced wastewater purification technology.
Chevron *Big Foot* – tension leg platform production facility (hull delivered March 2013; vessel completion estimated to be at end of 2013)	Daewoo Shipbuilding and Marine Engineering, Korea	$300 million (hull structure only)	373 feet	120,000 tons displacement	Upon completion, *Big Foot* will be the world's largest and deepest tension leg platform, with a capacity of 80,000 barrels of oil per day.
Ensco plc. DS-6 – drill ship (delivered January 2012)	Samsung Heavy Industries, Korea	$600 million	750 feet	105,822 tons displacement	Dynamically positioned drill ship can drill to 40,000 feet in up to 10,000 feet of water while maintaining a fixed, unanchored position.
Exxon Neftgas (a limited subsidiary of ExxonMobil) Sakhalin – 1 Project, Arkutun-Dagi Berkut Platform Topsides (estimated delivery June 2014)	Daewoo Shipbuilding and Marine Engineering, Korea	Proprietary information	Horizontal dimensions: 394 feet long by 230 feet wide Vertical dimensions: 328 feet from bottom of the deck to top of the drill rig	~40,000 tons displacement (topside unit only)	Upon completion in 2014, the topside will be transported and integrated to the offshore platform. The ice-resistant gravity based structure will become the largest oil and gas platform in Russia.
Noble Corporation *Noble Don Taylor* – drill ship (delivered April 2013)	Hyundai Heavy Industries, Korea	$600 million	752 feet	70,000 tons displacement	Dynamically positioned drill ship can drill to 40,000 feet in up to 10,000 feet of water while maintaining a fixed, unanchored position.

Table 7. (Continued)

Ship	Builder	Approximate cost (U.S. Dollars)	Length (feet)	Displacement (tons)	Notable characteristics
Norwegian Cruise Line, *Norwegian Gem* – cruise ship (delivered October 2007)	Meyer Werft, Germany	~$516 million	965 feet	50,259 tons displacement	The fourth ship in the *Jewel* ship class; has 1,188 cabins that can accommodate up to 2,384 passengers with a crew of 1,154.
Qatar Petroleum and ExxonMobil Q-Max – Liquefied Natural Gas (LNG) Carrier (lead ship delivered June 2008)	(1) Daewoo Shipbuilding and Marine Engineering, Korea (2) Samsung Heavy Industries, Korea	$300 million	1,132 feet	179,000 tons displacement	Revolutionary size for an LNG carrier with novel reliquefaction technology and first to use a two-rudder and propeller design. Ship class built using three different hull and structure designs.
Royal Caribbean Cruises, LTD. Oasis of the Sea – cruise ship (delivered October 2009)	STX Finland Turku Shipyard, Finland	$1.4 billion	1,187 feet	116,00 tons displacement	The largest cruise ship in the world, with a crew of 2,100. The ship can accommodate 5,408 passengers at double occupancy and incorporated advanced wastewater purification technology.
Seadrill Ltd. West Auriga – drill ship (delivered April 2013)	Samsung Heavy Industries, Korea	$600 million	748 feet	107,474 tons displacement	Dynamically positioned drill ship can drill to 37,000 feet in up to 12,000 feet of water while maintaining a fixed, unanchored position.
Star Deep Water Petroleum, LTD (a Chevron-affiliated company) Agbami – Floating Production, Storage and Offloading vessel (delivered June 2008)	Daewoo Shipbuilding and Marine Engineering, Korea	$1.2 billion (hull only)	1,049 feet	460,000 tons displacement (full load)	At the time of delivery was the world's largest FPSO, with a production capacity of 250,000 barrels of oil and 25 million cubic feet of natural gas per day.

Source: Industry-provided data.

APPENDIX IX: NAVY SHIPBUILDING CONTRACT TYPES

In Navy shipbuilding, the type of contract used can significantly influence the final cost of the ship. Table 8 below illustrates the basic differences between fixed-price and cost-reimbursement contracts and how each contract type can incentivize quality.

Table 8. Common Navy Shipbuilding Contract Types and Associated Risks to Quality Goals

Type of contract	Contract type use and application	Navy responsibility	Shipbuilder responsibility	Who assumes the risk of cost overruns	How quality is affected
Cost- reimbursement contracts with incentive fee	Used when: Requirements not well-defined or lack of knowledge does not permit costs to be sufficiently estimated to use a fixed-price contract. Applications: Commonly used on lead ships.	Pays contractor's allowable costs incurred, to the extent prescribed by the contract. Ship buyer is not guaranteed a completed ship at the expected level of quality within cost or schedule estimates.	Shipbuilder makes good faith effort to meet ship buyer's needs within the estimated cost.	Navy	Incentive fee may allow shipbuilder to earn higher fee if costs are kept low (e.g., by minimizing rework).
Fixed-price- incentive (firm target) contract	Used when: A ceiling price, target cost, target profit and profit adjustment formula can be established that will provide a fair and reasonable incentive. Provides for the contractor to assume an appropriate share of the risk. Applications: Commonly used for follow-on ships in a class.	Navy pays fixed target price which includes shipbuilder's profit, but agrees to share cost overruns (or underruns) up to a ceiling price.	Shipbuilder delivers a ship at the expected level of quality, meeting all requirements and specifications as specified in the contract at or below the ceiling price.	Shared risk between Navy and ship-builder up to agreed ceiling price. Shipbuilder generally bears most risk over that amount.	The Navy and the shipbuilder share cost overruns up to the agreed ceiling, which on previous contracts has been up to 138 percent of the target cost.

Table 8. (Continued)

Type of contract	Contract type use and application	Navy responsibility	Shipbuilder responsibility	Who assumes the risk of cost overruns	How quality is affected
Firm-fixed-price contract	Used when: Fair and reasonable prices can be established at the outset. Applications: Limited use on new construction auxiliary and support ships.	Pays fixed price even if actual total cost of the ship falls short of or exceeds the contract price.	Shipbuilder delivers a ship at the expected level of quality, meeting all requirements and specifications as specified in the contract.	Shipbuilder	Contract type assigns risk to the shipbuilder and may provide direct incentive to ensure timely delivery of the ship at the expected level of quality. Additional incentives for shipbuilder to limit risks to quality, improve production efficiencies, and reduce costs.

Source: GAO analysis of information obtained from the Federal Acquisition Regulation, and the Department of Defense's Contract Pricing Preference Guide.

End Notes

[1] See GAO, Best Practices: High Levels of Knowledge at Key Points Differentiate Commercial Shipbuilding from Navy Shipbuilding, GAO-09-322 (Washington, D.C.: May 13, 2009).

[2] A variety of terms are often used to indicate a work item is deficient, such as defect, nonconformance, corrective work or corrective action, and items included on a "punch" list (essentially a list of all remaining work and corrective work necessary to meet the ship buyer's requirements). Remaining work or incomplete work refers to work that has yet to be performed in order to make a finished product or previously identified deficiencies that have yet to be corrected.

[3] The American Society for Quality defines Six Sigma as a fact based, data driven philosophy of quality improvement that values defect prevention over defect detection by reducing variation and waste in a process.

[4] See appendix II, table 2 for a more detailed description of these phases.

[5] In some instances, commercial ships, such as liquefied natural gas carriers or drill ships, may undergo additional sea trials following delivery to test specific equipment related to the ships' missions and intended uses.

[6] On nuclear powered Navy ships, the ship's crew begins boarding and training prior to ship delivery.

[7] On non-nuclear ships, when deficiencies identified during the builder's sea trials are not resolved prior to acceptance trials, some of the uncorrected deficiencies are transferred and grouped together with the deficiencies INSURV inspectors identify during the acceptance trials process, if the inspectors determine that the deficiencies pertain to Navy operational capabilities, contracted requirements, and safety regulations.

[8] See GAOand, GAO, Defense Acquisitions: Improved Management Practices Could Help Minimize Cost Growth in Navy Shipbuilding Programs, GAO-05-183 (Washington, D.C.: Feb. 28, 2005).

[9] The International Maritime Organization is the United Nations' organization responsible for maritime affairs and develops international treaties in this area. For example, the Safety of Life at Sea Convention is generally regarded as the most important of all international treaties concerning the safety of commercial ships, and specifies minimum safety standards for the construction, equipment and operation of certain ships.

[10] In particular, ISO 9001 certification is commonly obtained by all types of manufacturing and production companies and can indicate to potential customers that a company has established and implements a defined minimum level of quality policies and standards.

[11] Navy officials stated that the certification requirement for submarines is as rigorous as that of classification societies. During the design, construction, and maintenance of submarines, they use an oversight and certification program analogous to ship classification—called Submarine Safety—as a means to help ensure requirements are met and a minimum level of safety is achieved. The officials noted the Submarine Safety program is a cornerstone of the Navy's submarine design, construction, and maintenance practices and provides the basis for certification of every submarine.

[12] The Military Sealift Command is largely comprised of civilian staff and mariners. Since 1960, Military Sealift Command-operated ships have obtained and maintained commercial ship classification certificates issued by ABS. Section 3316 of title 46 of the U.S. code (as amended), designates ABS as the agency for ship classification and related functions for U.S. Government-owned vessels.

[13] American Bureau of Shipping, The ABS Guide for Building and Classing Naval Vessels (2004).

[14] Since 2006, the Virginia class submarine program has also delivered multiple vessels. According to the Navy, the Virginia class program is considered to be one of the Navy's most well run programs, but we did not compare it to the surface ships due to differences in the delivery process.

[15] The Delivery on Completed Vessel clause also requires that prior to commencing acceptance trials the contractor has: (1) satisfactorily carried out builder's trials for which the contractor is responsible, (2) corrected all shipbuilder-responsible deficiencies discovered before completion of the builder's sea trials, unless otherwise agreed to by the contracting officer in writing; and, (3) corrected all shipbuilder-responsible deficiencies discovered after completion of the builder's sea trials which are determined by the contracting officer to be necessary to avoid an adverse effect on the operational capability of the ship.

[16] Appendixes III-VI provide data on the closing of deficiencies following delivery for selected surface ship classes where multiple ships were delivered to the Navy.

[17] Liquidated damages are amounts contractually stipulated as a reasonable estimation of actual damages to be recovered by one party if the other party breaches.

[18] The Federal Acquisition Regulation (FAR) authorizes use of a cost reimbursement contract when circumstances do not allow the agency to define its requirements sufficiently to allow for a fixed-price type of contract or uncertainties involved in contract performance do not permit costs to be estimated with sufficient accuracy to use any type of fixed-price contract. When using this contract type the government pays for all allowable incurred costs to the extent prescribed in the contract.

[19] The FAR authorizes use of fixed-price-incentive contracts when a firm fixed-price contract is not suitable; the nature of the supplies or services being acquired and other circumstances of the acquisition are such that the contractor's assumption of a degree of cost responsibility will provide a positive profit incentive for effective cost control and performance. If the contract also includes incentives on technical performance and/or

delivery, the performance requirements provide a reasonable opportunity for the incentives to have a meaningful impact on the contractor's management of the work. This contract type provides that the government and the contractor generally share costs— pursuant to a formula—above the target cost and below the ceiling price.

[20] In late 2011, DDG 108, DDG 109, DDG 111, and DDG 112 transitioned to firm-fixed- price contract line items from fixed-price-incentive contract line items. At the time only DDG 112 was still under construction.

[21] Payments made during construction of T-AGM 25 were based on the shipbuilder's achievement of certain shipbuilding milestones, such as keel laying.

[22] See GAO, Standards for Internal Control in the Federal Government, GAO/AIMD-00-21.3.1 (Washington, D.C.: Nov. 1999).

[23] Leading commercial ship buyers and shipbuilders in oil and gas, cruise and shipping industries referred to these plans by different terms, such as an inspection and test plan or yard quality standard.

[24] Nuclear submarines and aircraft carriers are also subject to additional oversight and inspection by NAVSEA 08 officials, who are responsible for nuclear propulsion systems. Navy submarines are subject to additional specialized inspections through its Submarine Safety program. Inspections conducted through this program, as well as audits and observations made by NAVSEA's Nuclear Propulsion Directorate of all shipyard activities, can also identify broader quality issues that affect construction. For example, Newport News Shipbuilding experienced issues with welders using incorrect filler material on Virginia Class Submarines. While Newport News was already working on the issue, audits by NAVSEA 08 identified it as a systemic problem, elevating the severity of the problem, according to officials. The Navy and the shipbuilder subsequently led a thorough review of weld filler materials used in potentially affected vessels.

[25] Officials at one other SUPSHIP location indicated their quality inspectors use the builder's production drawings when conducting inspections.

[26] We reviewed the quality assurance plans and policies of the eight Navy shipbuilding contractors and found them to be consistent with ISO 9000 quality management standards.

[27] The advanced technologies have been installed and tested on Royal Caribbean's Independence of the Seas and Liberty of the Seas cruise ships.

[28] ABS did grant a temporary classification certificate to LCS 1 for the sole purpose of transporting the ship from the shipbuilder's facility to the Port of Norfolk, Virginia.

[29] The Navy's LCS shipbuilding contracts require that the ships are built in accordance with the Naval Vessel Rules and other referenced ABS Rules and Guides as necessary to obtain classification. In the absence of ABS, the Navy intends to internally certify those aspects of the LCS designs that were not previously approved by the class society or were subsequently changed.

[30] Specifically, according to officials, the use of the phrase "to the satisfaction of the attending surveyor" in the applicable shipbuilding rule sets is meant to allow the marine surveyor the flexibility to use professional judgment as to whether or not a completed work item is structurally and mechanically suitable for the intended use and application. An item is "fit for service" if it can adequately operate as intended in its as-built condition.

[31] The Achieving Service Life Program is a cooperative agreement established in 2008 between ABS and NAVSEA. The program covers roughly 15-20 ships per year, focusing on those ships with dry-docking availabilities.

INDEX

Y

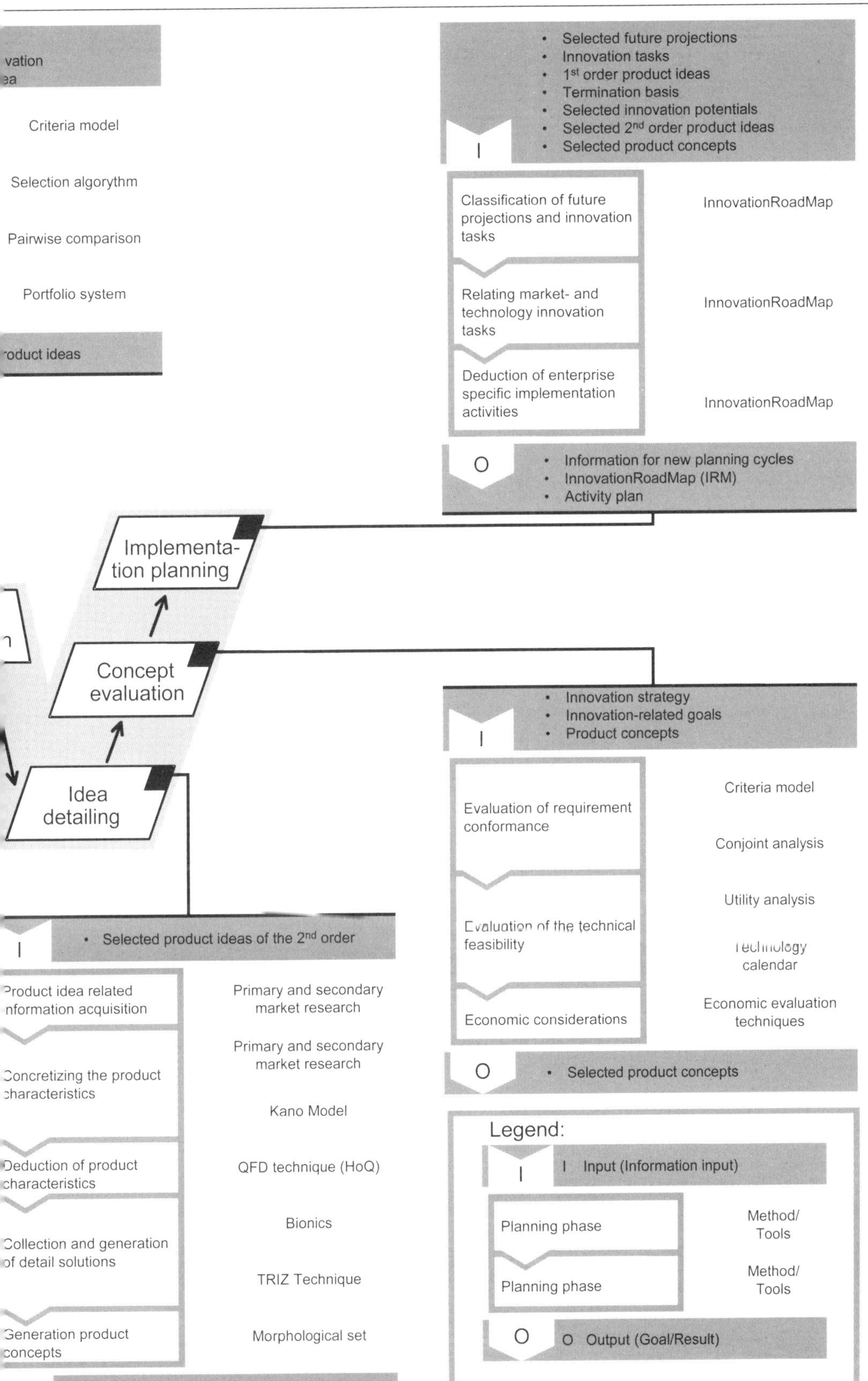

Cf. Brandenburg 2002